Lecture Notes in Artificial Intelligence   5395

Edited by R. Goebel, J. Siekmann, and W. Wahlster

Subseries of Lecture Notes in Computer Science

Lucas Paletta   John K. Tsotsos (Eds.)

Attention in Cognitive Systems

5th International Workshop
on Attention in Cognitive Systems, WAPCV 2008
Fira, Santorini, Greece, May 12, 2008
Revised Selected Papers

 Springer

Series Editors

Randy Goebel, University of Alberta, Edmonton, Canada
Jörg Siekmann, University of Saarland, Saarbrücken, Germany
Wolfgang Wahlster, DFKI and University of Saarland, Saarbrücken, Germany

Volume Editors

Lucas Paletta
Joanneum Research
Institute of Digital Image Processing
Wastiangasse 6, 8010 Graz, Austria
E-mail: lucas.paletta@joanneum.at

John K. Tsotsos
York University
Center for Vision Research (CVR)
and Department of Computer Science and Engineering
4700 Keele St., Toronto ON M3J 1P3, Canada
E-mail: tsotsos@cse.yorku.ca

Library of Congress Control Number: 2009921734

CR Subject Classification (1998): I.2, I.4, I.5, I.3, J.3

LNCS Sublibrary: SL 7 – Artificial Intelligence

ISSN 0302-9743
ISBN 978-3-642-00581-7 Springer Berlin Heidelberg New York

Typesetting: Camera-ready by author, data conversion by Scientific Publishing Services, Chennai, India
Printed on acid-free paper   SPIN: 12633487   06/3180   5 4 3 2 1 0

Preface

Attention has represented a core scientific topic in the design of AI-enabled systems in the last few decades. Today, in the ongoing debate, design, and computational modeling of artificial cognitive systems, attention has gained a central position as a focus of research. For instance, attentional methods are considered in investigating the interfacing of sensory and cognitive information processing, for the organization of behaviors, and for the understanding of individual and social cognition in infant development.

While visual cognition plays a central role in human perception, findings from neuroscience and experimental psychology have provided strong evidence about the perception–action nature of cognition. The embodied nature of sensory-motor intelligence requires a continuous and focused interplay between the control of motor activities and the interpretation of feedback from perceptual modalities. Decision making about the selection of information from the incoming sensory stream – in tune with contextual processing on a current task and an agent's global objectives – becomes a further challenging issue in attentional control. Attention must operate at interfaces between a bottom-up-driven world interpretation and top-down-driven information selection, thus acting at the core of artificial cognitive systems. These insights have already induced changes in AI-related disciplines, such as the design of behavior-based robot control and the computational modeling of animats.

Today, the development of enabling technologies such as autonomous robotic systems, miniaturized mobile – even wearable – sensors, and ambient intelligence systems involves the real-time analysis of enormous quantities of data. These data have to be processed in an intelligent way to provide "on time delivery" of the required relevant information. Knowledge has to be applied about what needs to be attended to, and when, and what to do in a meaningful sequence, in correspondence with visual feedback.

The individual contributions of this book meet these scientific and technological challenges on the design of attention and present the latest state of the art in related fields. The book evolved out of the 5th International Workshop on Attention in Cognitive Systems (WAPCV 2008) that was held on Santorini, Greece, as an associated workshop of the 6th International Conference on Computer Vision Systems (ICVS 2008). The goal of this workshop was to provide an interdisciplinary forum to examine computational models of attention in cognitive systems from an interdisciplinary viewpoint, with a focus on computer vision in relation to psychology, robotics and neuroscience. The workshop was held as a single-day, single-track event, consisting of high-quality podium and poster presentations. We received a total of 34 paper submissions for review, 22 of which were retained for presentations (13 oral presentations and 9 posters). We would like to thank the members of the Program Committee for their substantial contribution to the quality of the workshop. Two invited speakers strongly supported the success of the event with well-attended presentations given on "Learning to Attend: From Bottom-Up to Top-Down" (Jochen Triesch) and "Brain Mechanisms of Attentional Control" (Steve Yantis).

WAPCV 2008 and the editing of this collection was supported in part by The European Network for the Advancement of Artificial Cognitive Systems (euCognition). We are very thankful to David Vernon (co-ordinator of euCognition) and Colette Maloney of the European Commission's ICT Program on Cognition for their financial and moral support. Finally, we wish to thank Katrin Amlacher for her efforts in assembling these proceedings.

January 2009   Lucas Paletta
                                                                    John K. Tsotsos

Organization

Chairing Committee

Lucas Paletta   Joanneum Research, Austria
John K. Tsotsos   York University, Canada

Advisory Committee

Laurent Itti   University of Southern California, CA (USA)
Jan-Olof Eklundh   KTH (Sweden)

Program Committee

Leonardo Chelazzi   University of Verona, Italy
James J. Clark   McGill University, Canada
J.M. Findlay   Durham University, UK
Simone Frintrop   University of Bonn, Germany
Fred Hamker   University of Muenster, Germany
Dietmar Heinke   University of Birmingham, UK
Laurent Itti   University of Southern California, USA
Christof Koch   California Institute of Technology, USA
Ilona Kovacs   Budapest University of Technology, Hungary
Eileen Kowler   Rutgers University, USA
Michael Lindenbaum   Technion, Israel
Larry Manevitz   University of Haifa, Israel
Baerbel Martsching   University of Paderborn, Germany
Giorgio Metta   University of Genoa, Italy
Vidhay Navalpakkam   California Institute of Technology, USA
Aude Oliva   MIT, USA
Kevin O'Regan   Université de Paris 5, France
Fiora Pirri   University of Rome, La Sapienza, Italy
Marc Pomplun   University of Massachusetts, USA
Catherine Reed   University of Denver, USA
Ronald A. Rensink   University of British Columbia, Canada
Erich Rome   Fraunhofer IAIS, Germany
John G. Taylor   King's College London, UK
Jochen Triesch   Frankfurt Institute for Advanced Studies, Germany
Nuno Vasconcelos   University of California San Diego, USA
Chen Yu   University of Indiana, USA
Tom Ziemke   University of Skovde, Sweden

Sponsoring Institutions

euCognition - The European Network for the Advancement of Artificial Cognitive Systems
Joanneum Research, Austria

Table of Contents

Attention in Scene Exploration

On the Optimality of Spatial Attention for Object Detection 1
  Jonathan Harel and Christof Koch

Decoding What People See from Where They Look: Predicting Visual
Stimuli from Scanpaths 15
  Moran Cerf, Jonathan Harel, Alex Huth, Wolfgang Einhäuser, and
  Christof Koch

A Novel Hierarchical Framework for Object-Based Visual Attention 27
  Rebecca Marfil, Antonio Bandera, Juan Antonio Rodríguez, and
  Francisco Sandoval

Where Do We Grasp Objects? – An Experimental Verification of the
Selective Attention for Action Model (SAAM) 41
  Christoph Böhme and Dietmar Heinke

Contextual Cueing and Saliency

Integrating Visual Context and Object Detection within a Probabilistic
Framework 54
  Roland Perko, Christian Wojek, Bernt Schiele, and Aleš Leonardis

The Time Course of Attentional Guidance in Contextual Cueing 69
  Andrea Schankin and Anna Schubö

Conspicuity and Congruity in Change Detection 85
  Jean Underwood, Emma Templeman, and Geoffrey Underwood

Spatiotemporal Saliency

Spatiotemporal Saliency: Towards a Hierarchical Representation of
Visual Saliency 98
  Neil D.B. Bruce and John K. Tsotsos

Motion Saliency Maps from Spatiotemporal Filtering 112
  Anna Belardinelli, Fiora Pirri, and Andrea Carbone

Attentional Networks

Model Based Analysis of fMRI-Data: Applying the sSoTS Framework
to the Neural Basic of Preview Search 124
    Eirini Mavritsaki, Harriet Allen, and Glyn Humphreys

Modelling the Efficiencies and Interactions of Attentional Networks..... 139
    Fehmida Hussain and Sharon Wood

The JAMF Attention Modelling Framework 153
    Johannes Steger, Niklas Wilming, Felix Wolfsteller,
    Nicolas Höning, and Peter König

Attentional Modeling

Modeling Attention and Perceptual Grouping to Salient Objects 166
    Thomas Geerinck, Hichem Sahli, David Henderickx,
    Iris Vanhamel, and Valentin Enescu

Attention Mechanisms in the CHREST Cognitive Architecture......... 183
    Peter C.R. Lane, Fernand Gobet, and Richard Ll. Smith

Modeling the Interactions of Bottom-Up and Top-Down Guidance in
Visual Attention 197
    David Henderickx, Kathleen Maetens, Thomas Geerinck, and
    Eric Soetens

Relative Influence of Bottom-Up and Top-Down Attention 212
    Matei Mancas

Towards Standardization of Evaluation Metrics and Methods for Visual
Attention Models 227
    Muhammad Zaheer Aziz and Bärbel Mertsching

Comparing Learning Attention Control in Perceptual and Decision
Space 242
    Maryam S. Mirian, Majid Nili Ahmadabadi, Babak N. Araabi, and
    Ronald R. Siegwart

Automated Visual Attention Manipulation 257
    Tibor Bosse, Rianne van Lambalgen, Peter-Paul van Maanen, and
    Jan Treur

Author Index 273

On the Optimality of Spatial Attention for Object Detection

Jonathan Harel and Christof Koch

California Institute of Technology, Pasadena, CA, 91125

Abstract. Studies on visual attention traditionally focus on its physiological and psychophysical nature [16,18,19], or its algorithmic applications [1,9,21]. We here develop a simple, formal mathematical model of the advantage of spatial attention for object detection, in which spatial attention is defined as processing a subset of the visual input, and detection is an abstraction with certain failure characteristics. We demonstrate that it is suboptimal to process the entire visual input given prior information about target locations, which in practice is almost always available in a video setting due to tracking, motion, or saliency. This argues for an attentional strategy independent of computational savings: no matter how much computational power is available, it is in principle better to dedicate it preferentially to selected portions of the scene. This suggests, anecdotally, a form of environmental pressure for the evolution of foveated photoreceptor densities in the retina. It also offers a general justification for the use of spatial attention in machine vision.

1 Introduction

Most animals with visual systems have evolved the peculiar trait of processing subsets of the visual input at higher bandwidth (faster reaction times, lower error rates, higher SNR). This strategy is known as focal or spatial attention and is closely linked to sensory (receptor distribution in the retina) and motor (eye movements) factors. Motivated by such wide-spread attentional processing, many machine vision scientists have developed computational models of visual attention, with some treating it broadly as a hierarchical narrowing of possibilities [1,2,8,9,17]. Several studies have demonstrated experimental paradigms in which various such attentional schemes are combined with recognition/detection algorithms, and have documented the resulting computational savings and/or improved accuracy [4,5,6,7,20,21].

Here, we seek to describe a general justification for spatial attention in the context of an object detection goal (detecting targets in images wherever they occur). We take an abstract approach to this phenomenon, in which both the attentional and detection mechanisms are independent of the conclusions. Similar frameworks have been proposed by other authors [3,10]. The most common justification for attentional processing, in particular in visual psychology, is the computational saving that accrue if processing is restricted to a subset of the image. For machine vision scientists, in an age of ever decreasing computational

L. Paletta and J.K. Tsotsos (Eds.): WAPCV 2008, LNAI 5395, pp. 1–14, 2009.

costs of digital processors, and for biologists in general, the question is whether there are other justifications for the *spatial spotlight of attention*. We will address this in three steps which form the core substance of this paper:

1. (Section 2) We demonstrate that object detection accuracy can be improved using attentional selection in a motivating machine vision experiment.

2. (Section 3) We model *a generalized form* of this system and demonstrate that accuracy is optimal with attentional selection if prior information about target locations is not or cannot be used to bias detector output.

3. (Section 4) We then demonstrate that, even if priors are used optimally, if there is a fixed computational resource which can be concentrated or diluted over locations in the visual scene, with corresponding modulations in accuracy, that it is optimal to process only the most likely target locations. We show how the optimal extent of this spatial attention depends on the environment, quantified as a specific tolerance for false positives and negatives.

2 Motivating Example

2.1 Experiment

An important problem in machine vision is the detection of objects from broad categories in cluttered scenes, in which a target may only take up a small fraction of the available pixels. We built a system to solve an instance of this "object detection" problem: detecting cars and pedestrians wherever they occurred in a fully annotated video of 4428 frames, captured at 15fps at VGA (640x480) resolution.

Training images (47,459 total, of which 4,957 are positive examples) were gathered from [11] and [12]. The object detection system worked in two steps for each frame independently:

1. A saliency heat map [9] for the frame (consisting of color, orientation, intensity, motion, and flicker channels) was computed and subsequently serialized into an ordered list of "fixation" locations (points) using a choose-maximum/inhibit-its-surround iterative loop. A rectangular image crop ("window") around each fixation location was selected using a crude flooding-based segmentation algorithm.

2. The first $F \in \{1, 3, 5, 7, 9\}$ fixation windows were then processed using a detection module (one for cars and one for pedestrians), which in turn decided if each window contained its target object type or not. The detection modules based their classification decision on the output of an SVM, with input vectors having components proportional to the multiplicity of certain quantized SIFT [14] features over an image subregion, with subregions forming a pyramid over the input image – this method has proven quite robust on standard benchmarks [13].

2.2 Results

We quantified the performance by recording four quantities for each choice of F windows per frame: (1) True Positive Count (TPC) – the number of windows,

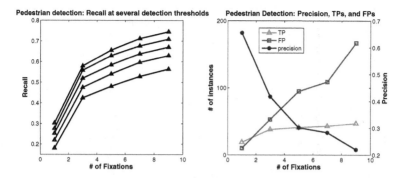

Fig. 1. Result of running detector over entire video. As the number of windows processed per frame increases, recall rate increases (left), while precision rate decreases (right). Left: curves for different settings of the SVM detection threshold.

pooled over the entire video[1], in which a detection corresponded to a true object at that location. (2) False Positive Count (FPC) – windows labeled as a target where there was actually not one, and using the False Negative Count, FNC (number of targets undetected), (3) precision = TPC/(TPC+FPC) – fraction of detections which were actually target objects, and (4) recall = TPC/(TPC+FNC) – fraction of target objects which were detected.

The results for pedestrian detection are shown in Fig. 1. Results on cars were qualitatively equivalent.

Each data point in Fig. 1 corresponds to results over the pooled video frames, but at each frame the number of windows processed is not the same: we parameterize over this window count along the x-axis. All plots in this paper use this underlying *attention-parameterizing* scheme, in which processing one window corresponds to maximally focused attention, and processing them all corresponds to maximally blurred attention. The results in Fig. 1 indicate that, in our experiment, the recall rate increases as more windows are processed per frame, whereas the precision rate falls off. Therefore, in this case, it is reasonable to process just a few windows per frame, i.e., implement an attentional focus, in order to balance performance, independent of computational savings.

This can be understood by considering that lower-saliency windows are a priori unlikely to contain a target, and so their continued processing yields a false positive count that accumulates at nearly the false positive rate of the detector. The true positive count, on the other hand, saturates at a small number proportional to the number of targets in the scene. These two trends yield a decreasing precision ratio. This is seen more directly in Fig. 2 below, where we plot the average number of pedestrians contained in the first F fixation windows of a frame, noting that the incremental increase (slope) per added window is decreasing. We will see in the next section how the behavior observed here is sensitive to incorporating priors into detection decisions.

---

[1] Results shown are for 20% of the frames uniformly sampled from the video.

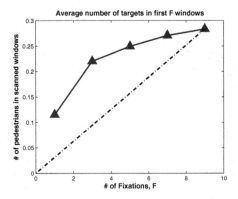

Fig. 2. The average number of pedestrians contained in the first F windows. The dotted line connects the origin to the maximum point on the curve, showing what we would observe if pedestrians were equally likely to occur at each fixation. But since targets are more likely to occur in early fixations, the slope decreases.

3   A Simple Mathematical Model of Spatial Attention for Object Detection

In this section, we model *a generalized form* of the system in the experiment above, and explore its behavior and underlying assumptions.

3.1   Preliminaries

We suppose henceforth that our world consists of images/frames streaming into our system, that we form window sets over these images, *somehow* sort these windows in a negligibly cheap way (e.g., according to fixation order from a saliency map, or due to an object tracking algorithm), and then run an object detection module (e.g., a pedestrian detector) over only the first w of these windows on each frame, according to sorted order, where $w \in \{1, 2, ..., N\}$. We will refer to the processing of only the first w windows as *spatial attention*, and the smallness of w as the *extent of spatial attention*.[2]

We will model the behavior of a detection system as a function of w. Define[3]

$T(w) \doteq \#$ targets in first w windows

$FPC(w) \doteq \#$ false positives in first w windows (incorrect detections)

$TPC(w) \doteq \#$ true positives in first w windows (correct detections)

$FNC(w) \doteq \#$ false negatives (in entire image after processing w windows)

$TNC(w) \doteq \#$ true negatives (in entire image after processing w windows)

These counts determine the performance of the detection system, and so we will calculate their expected values, averaged over many frames. To do this, we define

---

[2] See Appendix for table of parameters.

[3] C is for count, as in FalsePositiveCount = FPC.

the following: For a single frame/image, let T_i be the binary random variable indicating whether there is in truth a target at window i, with 1 corresponding to presence. Let D_i be the binary random variable indicating the result of the detection on window i, with 1 indicating a detection. Then:

$$E[T(w)] = \sum_{i=1}^{w} E[T_i] = \sum_{i=1}^{w} p_i, \text{ where } p_i \doteq \Pr\{T_i = 1\}$$

$$E[FPC[w]] = \sum_{i=1}^{w} E[FP_i] \text{ where } FP_i = \begin{cases} 1 \text{ if } D_i = 1 \text{ and } T_i = 0 \\ 0 \qquad\qquad \text{otherwise} \end{cases}$$

$$= \sum_{i=1}^{w} p(D_i = 1 | T_i = 0) \cdot (1 - p_i) = fpr \cdot (w - E[T(w)])$$

$$E[TPC(w)] = \sum_{i=1}^{w} E[TP_i], \text{ where } TP_i = \begin{cases} 1 \text{ if } D_i = 1 \text{ and } T_i = 1 \\ 0 \qquad\qquad \text{otherwise} \end{cases}$$

$$= \sum_{i=1}^{w} p(D_i = 1 | T_i = 0) \cdot p_i = tpr \cdot E[T(w)]$$

Where the false and true positive rates, $fpr \doteq p(D_i = 1 | T_i = 0) \; \forall i$, and $tpr \doteq p(D_i = 1 | T_i = 1) \; \forall i$, are taken to be properties of the detector. Similarly,

$$E[FNC(w)] = n - E[TPC(w)], \text{ where } n \doteq E\left[\sum_{i=1}^{N} T_i\right] = \sum_{i=1}^{N} p_i = E[T(N)]$$

Since $\sum_{i=1}^{N} T_i = TPC(w) + FNC(w) = \#$ of windows with a target in image

And

$$E[TNC(w)] = (N - n) - E[FPC(w)], \text{ because:}$$

$$N - \sum_{i=1}^{N} T_i = FPC(w) + TNC(w) = \# \text{ of windows without a target in image.}$$

3.2 Decreasing Precision Underlies Utility of Spatial Attention

We shall now use the quantities defined above to model the precision and recall trends demonstrated in the motivating example. But, first we must make a modeling assumption: suppose that p_i is decreasing in i such that:

$$E[T(w)] = n \frac{1 - \exp(-w/k)}{1 - \exp(-N/k)} \tag{1}$$

which has a similar form to that in Fig. 2. Note that this yields $E[T(0)] = 0$, and $E[T(N)] = n$, as above, where n represents the average number of target-containing windows in a frame. In Fig. 3, we plot this profile for several settings

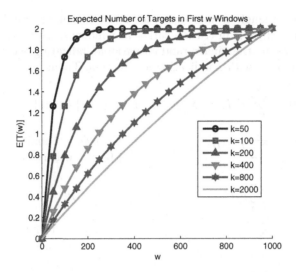

Fig. 3. A model of the average number of targets in highest w priority windows

of k, with $n = 2$ and $N = 1000$ (more nearly continuous/graded than the motivating experiment for smoothness).

Larger values of k correspond to $E[T(w)]$ profiles which are closer to linear. Linearly increasing $E[T(w)]$ corresponds to constant p_i so that $\sum_{i=1}^{w} p_i$ increases an equal amount for each increment of w. Concave down profiles above the line corresponding to decreasing p_i profiles, in which the incremental contribution to $E[T(w)]$ from $\sum_{i=1}^{w} p_i$ is higher for low w. Such decreasing p_i represent an ordering of windows where early windows are more likely to contain targets than later windows. In practice, one can almost always arrange such an ordering since targets are likely to remain in similar locations from frame to frame, be salient, or move, or be a certain color, etc.. Here, we are not concerned with how this ordering is carried out, but assume that it is.

Let subscript-M denote a particular count accumulated over M frames. As the number of frames M grows,

$$\lim_{M \to \infty} T_M(w) = \lim_{M \to \infty} \sum_{image=1}^{M} T_{image}(w) = M \cdot E[T(w)]$$

by the Central Limit Theorem, where $T_{image}(w)$ is the number of targets in *image*. Using similar notation, the precision after M images have been processed approaches:

$$\lim_{M \to \infty} prec_M(w) = \lim_{M \to \infty} \frac{TPC_M(w)}{TPC_M(w) + FPC_M(w)}$$

$$= \frac{M \cdot E[TPC(w)]}{M \cdot E[TPC(w)] + M \cdot E[FPC_M(w)]} = \frac{E[TPC(w)]}{E[TPC(w)] + E[FPC_M(w)]}$$

Equivalently, the recall approaches

$$\lim_{M \to \infty} rec_M(w) = \frac{E[TPC(w)]}{E[TPC(w)] + E[FNC_M(w)]}.$$

Define $prec(w) \doteq \lim_{M \to \infty} prec_M(w)$, and $rec(w) \doteq \lim_{M \to \infty} rec_M(w)$.

Using the model equation (1), and the equilibrium precision and recall definitions, we see that we can qualitatively reproduce the experimental results observed in Fig. 1, as seen in Fig. 4.

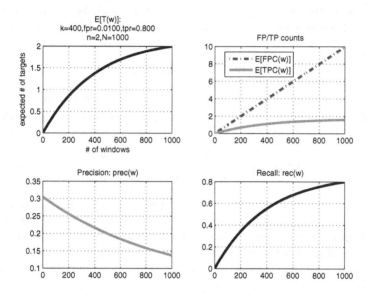

Fig. 4. Equilibrium precision and recall rates using a model $E[T(w)]$

Simulation results suggest that this decreasing precision, increasing recall holds under a wide variety of concave profiles $E[T(w)]$ (including all parameterized in (1)), and detector rates properties (tpr, fpr). A few degenerate cases will flatten the precision curve: a linear $E[T(w)]$ and/or a zero false positive rate, i.e., zero ability to order windows, and a perfect detector, respectively. Otherwise, recall and precision pull performance in opposite directions over the range of w, and optimal performance will be somewhere in the middle depending on the exact parameters and objective function, e.g., area under ROC or precision-recall curve. Therefore, it is in this context best to process only the windows most likely to contain a target in each frame, i.e., implement a form of spatial attention.

tpr, fpr fixed $\forall i$ means having little faith in, or no ability to calculate, one's prior belief. This model is realistic if one does not have faith in, or ability to calculate, one's prior belief: i.e., the order of windows is known, but not specifically $P(T_i = 1)$. Formally, in a Bayesian setting, one would assume that there is a pre-decision detector output $D_{ic} \in \boldsymbol{\theta}$ with constant known densities $p(D_{ic}|T_i)$. Then,

$$tpr = P(D_i = 1|T_i = 1) = \Pr(D_{ic} \in \theta^+|T_i = 1), \qquad (2)$$

where θ^+ is the largest set such that

$$LLR = \frac{p(D_{ic}|T_i = 1)P(T_i = 1)}{p(D_{ic}|T_i = 0)P(T_i = 0)} > 1 \ \forall D_{ic} \in \theta^+ \qquad (3)$$

Very notably, the definition in (2) yields a tpr which is *not the same for all* i (as modeled previously), and in particular, which depends on the prior $P(T_i = 1) = p_i$. Similarly,

$$fpr = P(D_i = 1|T_i = 0) = \Pr(D_{ic} \in \theta^+|T_i = 0),$$

also depends on p_i. Only if one assumes that $P(T_i = 1) = P(T_i = 0)$, then (3) is the same for all i, and so is (2). Having constant tpr and fpr $\forall i$ is also equivalent to evaluating the likelihood ratio as:

$$LLR = \left(\frac{p(D_{ic}|T_i = 1)}{p(D_{ic}|T_i = 0)}\right)^\gamma \frac{P(T_i = 1)}{P(T_i = 0)}$$

in the limit as $\gamma \to \infty$, or putting little faith into the prior distribution. This is somewhat reasonable given the motivating experimental example in section 2. The output of the detector is somehow much more reliable than whether a location was salient in determining the presence of a target, and the connection between saliency and probability of a target $P(T_i = 1)$ may be changing or incalculable.

Importantly, if a prior distribution is available explicitly, then the false positive counts $FPC(w)$ saturate at high values of w which are unlikely to contain a target, and the utility of not running the detector on some windows is eliminated, although it still saves compute cycles.

4   Distributing a Fixed Computational Resource

In the previous section, we assume that it makes sense to process a varying number of windows with the same underlying detector for each window. A more realistic assumption about systems in general is that they have a fixed computational resource, and that it can be and should be fully used to transform input data into meaningful detector outputs.

Now, suppose the same underlying two-step model as before: frames of images stream in to our system, we somehow cheaply generate an ordered window set on each of these, and select a number w of the highest-priority windows, each of which will pass through a detector.

Here, we impose an additional assumption: that the more detection computations are made (equivalently, the more detector instances there are to run in parallel), the weaker each individual detection computation/detector must be, in accordance with the conservation of computational resource. Below, we derive a simple detector degradation curve, and then use it to characterize the relationship between the risk priorities of a system (tolerance for false positives/negatives) and its optimal extent of spatial attention, viz., how many windows down the list it should analyze.

4.1 More Detectors, Weaker Detectors

We assume that a detector DT is an abstraction which provides us with information about a target. For simplicity, suppose that it informs us about a particular real-valued target property x, like its automobility or pedestrianality. Then the information provided by detector DT is:

$$I_{DT} \doteq H_0 - H_{DT} \doteq H(P_0(x)) - H(P_{DT}(x))$$

where $P_{DT}(x)$ is the density function over x output by the detector, and $H_0 = H(P_0(x))$ is the entropy in x before detection, where $P_0(x)$ is the prior distribution over x.

It seems intuitively clear that given fixed resources, one can get more information out of an aggregate of cheap detectors than out fewer more expensive detectors. One way to quantify this is by assuming that the fixed computational resource is the number of compute "neurons" R, and that these neurons can be allocated to understanding/detecting in just one window, or divided up into s sets of R/s neurons, each of which will process a different window/spatial location. There are biological data suggesting that neurons from primary sensory cortices to MTL [15] fire to one concept/category out of a set, i.e. that the number of concepts encodable with n neurons is roughly proportional to n, and so the information n neurons carry is proportional to $\log(n)$. Thus, a good model for how much information each of s detectors provides is $\log\left(\frac{R}{s}\right)$, where $\log(R)$ is some constant amount of information provided if the entire computational resource were allocated to one detector.

Let DT_1 denote the singleton detector comprised of using the entire computational resource R, and DT_s denote one of the s detectors using only R/s "neuronal" computational units. Then,

$$I_{DT_1} = H_0 - H_{DT_1} = \log(R), \text{ and } I_{DT_s} = H_0 - H_{DT_s} = \log(R/s) \iff$$
$$H_{DTs} - H_{DT_1} = \log(R) - \log(R/s) = \log(s),$$

that is, that the output of each of s detectors has $\log(s)$ bits more uncertainty in it than the singleton detector.

4.2 FPC, TPC, FNC, and TNC for This System

We will assume this time that the detector is Bayes optimal, i.e. that it incorporates the prior information into its decision threshold. For simplicity, and with some loss of generality, assume that the output probability density on x

of the detectors is Gaussian around means $+1$ and -1 corresponding to target present and absent, resp., with standard deviation σ_{DT}. Then, since the differential entropy of a Gaussian is $\log(\sigma\sqrt{2\pi e})$, a distribution which is $\log(s)$ bits more entropic than the normal with σ_{DT_1} has standard deviation $s \cdot \sigma_{DT_1}$, where σ_{DT_1} characterizes the output density over x of the detector which uses the entire computational resource. Therefore, since we assume we process w windows, we will employ detectors with output distributions having $\sigma = w \cdot \sigma_{DT_1}$.

The expected false positive count of our system, if it examines w windows is, from section 3.1:

$$E[FPC(w)] = \sum_{i=1}^{w} p(D_i = 1|T_i = 0)p(T_i = 0)$$

$$= \sum_{i=1}^{w} fpr_i \cdot p(T_i = 0) \tag{4}$$

To calculate fpr_i, we examine the likelihood ratio at window i, corresponding to the prior p_i :

$$LLR_i = \frac{p(D_i|T_i = 1)}{p(D_i|T_i = 0)} \frac{p_i}{1 - pi}$$

$$= \frac{\exp(-(D_i - 1)^2/2\sigma^2)}{\exp(-(D_i + 1)^2/2\sigma^2)} \cdot \frac{p_i}{1 - pi}$$

$$= \exp(2D_i/\sigma^2) \cdot \frac{p_i}{1 - pi}$$

$D_i = 1$ when $LLR_i > 1 \Longrightarrow$

$$\exp(2D_i/\sigma^2) > \frac{1 - p_i}{p_i} \Longleftrightarrow 2D_i/\sigma^2 > \log\left(\frac{1 - p_i}{p_i}\right) \Longleftrightarrow$$

$$D_i > \frac{\sigma^2}{2} \log\left(\frac{1 - p_i}{p_i}\right)$$

Thus,

$$fpr_i = p\left(D_i > \frac{\sigma^2}{2} \log\left(\frac{1 - p_i}{p_i}\right)\Big| T_i = 0\right)$$

$$= Q\left(\frac{\frac{\sigma^2}{2} \log\left(\frac{1-p_i}{p_i}\right) + 1}{\sigma}\right)$$

$$= Q\left(\frac{\sigma}{2} \log\left(\frac{1 - p_i}{p_i}\right) + \frac{1}{\sigma}\right) \tag{5}$$

where $Q(\cdot)$ is the complementary cumulative distribution function the standard normal. Substituting (5) into (4) gives:

$$E[FPC(w)] = \sum_{i=1}^{w} Q\left(\frac{\sigma}{2} \log\left(\frac{1 - p_i}{p_i}\right) + \frac{1}{\sigma}\right)(1 - p_i). \tag{6}$$

Similarly,

$$E[TPC(w)] = \sum_{i=1}^{w} Q\left(\frac{\sigma}{2}\log\left(\frac{1-p_i}{p_i}\right) - \frac{1}{\sigma}\right) p_i \qquad (7)$$

and the other two are dependent on these as usual: $E[TNC(w)] = (N-n) - E[FPC(w)]$, and $E[FNC(w)] = n - E[TPC(w)]$.

4.3 Optimal Distributions of the Computational Resource

Equations (6)-(7) are difficult to analyze as a function of w analytically, so we investigate their implications numerically. To begin, we use a model from equation (1), with $n = 3$ expected targets per total frame, $N = 100$ windows, prior profile parameter $k = 20$, and $\sigma_{DT_1} = 2/N$. The results are shown in Fig. 5.

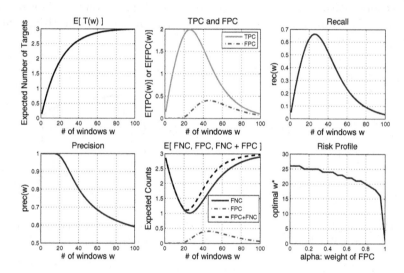

Fig. 5. Performance of an object detection system with fixed computational resource

We observe the increasing recall, decreasing precision trend for low w values, now *even with perfect knowledge of the prior*. This suggests that, at least for this setting of parameters, resources are best concentrated among just a few windows. The most striking feature of these plots, for example of the expected true positive count shown in green, is that there is an optimum around 20 or so windows. This corresponds to where the aggregate information of the thresholded detectors is peaked – beyond that, the detectors are spread too thinly and become less useful. Note that this is in contrast to the aggregate information of the pre-threshold real-valued detection outputs, which increases monotonically as $w \log(R/w)$.

It is interesting to understand not only that subselecting visual regions is beneficial for performance, but how the exact level of spatial attention depends on other factors. We now introduce the notion of a "Risk Profile":

$$w^*(\alpha) = \arg\min_{w} \{\alpha E[FPC(w)] + (1 - \alpha)E[FNC(w)]\}.$$

That is, suppose a system has penalty function which depends on the false positives and false negatives. Both should be small, but how the two compare might depend on the environment: a prey may care a lot more about false negatives than a predator, e.g.. For a given false positive weight α, the optimal w^* corresponds to number of windows among which the fixed computational resource should be distributed in order to minimize penalty. We find (see Fig. 6), that an increasing emphasis on false negatives (low α), leads to a more thinly distributed attentional resource being optimal. Thus, in light of this simple analysis, it makes sense that an animal with severe false negative penalties, such as a grazer with wolves on the horizon, may have evolved to spread out its sensory-cortical hardware over a larger spatial region – and indeed grazers have an elongated visual streak rather than a small fovea.

The general features of the plots shown in Fig. 5 hold over a wide range of parameters. We summarize the numerical findings by showing the risk profiles for a few such parameter ranges in Fig. 6.

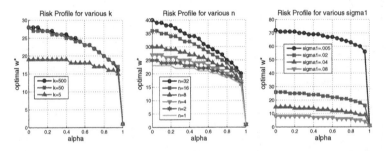

Fig. 6. The optimal number of windows out of 100 to process, for increasing α, the importance of avoiding false positives relative to false negatives. $sigma1 \equiv \sigma_{DT_1}$.

The important feature of all these plots is that the optimal number of windows w over which to distribute computation in order to minimize the penalty function is always less than $N = 100$, and that the risk profiles increase to the left, with increasing false negative count importance, for a wide range of parameterized conditions.

5   Conclusions

We have demonstrated, first in experiment and then using a simple numerical model, the critical importance of attentional selection for increased accuracy in a detection task. We find that processing scene portions which are a priori unlikely to contain a target can hurt performance if this prior information is not utilized to bias detection decisions. However, if the computational resources available for detection are fixed and must be distributed somehow to various scene portions, with a corresponding dilution in accuracy, it is best to concentrate them on scene portions which are a priori likely to contain a target, even if prior information biases detector outputs optimally. Note that this argues for an attentional strategy independent of computational savings – no matter how great the computational resource, it is best focused attentionally. We also show how a system which prioritizes false negatives high relative to false positives benefits from a blurred focus of attention, which may anecdotally suggest an evolutionary pressure for the variety in photoreceptor distributions in the retinae of various species. In conclusion, we provide a novel framework within which to understand the utility of spatial attention, not just as an efficiency heuristic, but as fundamental to object detection performance.

Acknowledgements

We wish to thank DARPA for its generous support of a research program for the development of a biologically modeled object recognition system, and our close collaborators on that program, Sharat Chikkerur at MIT, and Rob Peters at USC.

References

1. Tsotsos, J.K., Culhane, S.M., Kei Wai, W.Y., Lai, Y., Davis, N.: Modeling visual attention via selective tuning. Artificial Intelligence (1995)
2. Amit, Y., Geman, D.: A Computational Model for Visual Selection. Neural Computation (1999)
3. Yu, A., Dayan, P.: Inference, Attention, and Decision in a Bayesian Neural Architecture. In: Proc. Neural Information Processing Systems (NIPS) (2004)
4. Bonaiuto, J., Itti, L.: Combining attention and recognition for rapid scene analysis. In: Proc. IEEE-CVPR Workshop on Attention and Performance in Computer Vision (WAPCV 2005) (2005)
5. Rutishauser, U., Walther, D., Koch, C., Perona, P.: Is attention useful for object recognition? In: Proc. International Conference on Computer Vision and Pattern Recognition (CVPR) (2004)
6. Miau, F., Papageorgiou, C.S., Itti, L.: Neuromorphic algorithms for computer vision and attention. In: Proceedings of Annual International Symposium on Optical Science and Technology (SPIE) (2001)
7. Moosmann, F., Larlus, D., Jurie, F.: Learning Saliency Maps for Object Categorization. In: ECCV International Workshop on The Representation and Use of Prior Knowledge in Vision (2006)
8. Koch, C., Ullman, S.: Shifts in selective visual attention: towards the underlying neural circuitry. Hum. Neurobiol. (1985)
9. Itti, L., Koch, C.: Computational modeling of visual attention. Nature Reviews Neuroscience (2001)
10. Ye, Y., Tsotos, J.K.: Where to Look Next in 3D Object Search. In: Proc. of Internat. Symp. on Comp. Vis. (1995)
11. http://cbcl.mit.edu/software-datasets/streetscenes/
12. http://labelme.csail.mit.edu/

13. Lazebnik, S., Schmid, C., Ponce, J.: Beyond bags of features: Spatial pyramid matching for recognizing natural scene categories. In: Proc. IEEE Conference on Computer Vision and Pattern Recognition (CVPR) (2006)
14. Lowe, D.G.: Distinctive Image Features from Scale-Invariant Keypoints. International Journal of Computer Vision (2004)
15. Waydo, S., Kraskov, A., Quian Quiroga, R., Fried, I., Koch, C.: Sparse Representation in the Human Medial Temporal Lobe. Journal of Neuroscience (2006)
16. Treisman, A.: How the deployment of attention determines what we see. Visual Cognition (2006)
17. Viola, P., Jones, M.: Rapid object detection using a boosted cascade of simple features. In: Proc. Computer Vision and Pattern Recognition (CVPR)(2001)
18. Pashler, H.E.: The Psychology of Attention. MIT Press, Cambridge (1998)
19. Braun, J., Koch, C., Davis, J.L. (eds.): Visual Attention and Cortical Circuits. MIT Press, Cambridge (2001)
20. Walther, D., Koch, C.: Modeling attention to salient proto-objects. Neural Networks (2006)
21. Mitri, S., Frintrop, S., Pervolz, K., Surmann, H., Nuchter, A.: Robust Object Detection at Regions of Interest with an Application in Ball Recognition. In: Proc. of International Conference on Robotics and Automation (ICRA) (2005)

Appendix

Table of parameters:

N	# of windows available to process in a frame
w	# of windows processed in a frame
n	average # of target-containing windows in a frame
k	poverty of prior information $\Rightarrow$ lower k, better a priori sorting of windows
σ_{DT_1}	standard deviation of detector output, if only one detector is used

Decoding What People See from Where They Look: Predicting Visual Stimuli from Scanpaths

Moran Cerf[1,*,**], Jonathan Harel[1,*], Alex Huth[1], Wolfgang Einhäuser[2], and Christof Koch[1]

[1] California Institute of Technology, Pasadena, CA, USA
moran@klab.caltech.edu
[2] Philipps-University Marburg, Germany

Abstract. Saliency algorithms are applied to correlate with the overt attentional shifts, corresponding to eye movements, made by observers viewing an image. In this study, we investigated if saliency maps could be used to predict which image observers were viewing given only scanpath data. The results were strong: in an experiment with 441 trials, each consisting of 2 images with scanpath data - pooled over 9 subjects - belonging to one unknown image in the set, in 304 trials (69%) the correct image was selected, a fraction significantly above chance, but much lower than the correctness rate achieved using scanpaths from individual subjects, which was 82.4%. This leads us to propose a new metric for quantifying the importance of saliency map features, based on discriminability between images, as well as a new method for comparing present saliency map efficacy metrics. This has potential application for other kinds of predictions, e.g., categories of image content, or even subject class.

1 Introduction

In electrophysiological studies, the ultimate validation of the relationship between physiology and behavior is the decoding of behavior from physiological data alone [1,2,3,4,5,6,7]. If one can determine which image an observer has seen using only the firing rate of a single neuron, one can conclude that that neuron's output is highly informative about the image set. In psychophysical studies it is common to show an observer (animal or human) a sequence of images or video while recording their eye movements using an eye-tracker. Often, such studies aim to predict subjects' scanpaths using saliency maps [8,9,10,11], or other techniques [12,13]. The predictive power of a saliency model is typically judged by computing some similarity metric between scanpaths and the saliency map generated by the model [8,14]. Several similarity metrics have become de facto standards, including NSS [15] and ROC [16]. A principled way to assess the goodness of such a metric is to compare its value for scanpath-saliency map pairs which correspond to the same image and different images. If this difference

---

* These authors contributed equally to this work.
** Corresponding author.

is systematic, one can apply the metric to several candidate saliency maps per image, and asses which saliency map yields the highest decodability.

This decodability represents a new measure of saliency map efficacy. It is complementary to the current approaches: rather than predicting fixations from image statistics, it predicts image content from fixation statistics. The fundamental advantage of rating saliency maps in this way is that the score reflects not only how similar the scanpath is to the map, but also how *dissimilar it is from the maps of other images*. Without that comparison, it is possible to artificially inflate similarity metrics using saliency heuristics which increase the correlation with all scanpaths, rather than only those recorded on the corresponding image. Thus, we propose this as an alternative to the present measures of saliency maps' predictive power, and test this on established eye-tracking datasets.

The contributions of this study are:

1. A novel method for quantifying the goodness of an attention prediction model based on the stimuli presented and the behavior.
2. Quantitative results using this method that rank the importance of feature maps based on their contribution to the prediction.

2   Methods

2.1   Experimental Setup

In order to test if scanpaths could be used to predict which image from a set was being observed at the time it was recorded, we collected a large dataset of images and scanpaths from various earlier experiments (from the database of [17]). In all of these previous experiments, images were presented to subjects for 2 s, after which they were instructed to answer "How interesting was the image?" on a scale of 1-9 (9 being the most interesting). Subjects were not instructed to look at anything in particular; their only task was to rate the entire image. Subjects were always naïve to the purpose of the experiments. The subset of images was presented for each subject in random order.

Scenes were indoors and outdoors still images (see examples in Fig. 1), containing faces and objects. Faces were in various skin colors and age groups, and exhibiting neutral expressions. The images were specifically composed so that the faces and objects appeared in a variety of locations but never in the center of the image, as this was the location of the starting fixation on each image. Faces and objects vary in size. The average size was $5\% \pm 1\%$ (mean $\pm$ s.d.) of the entire image - between $1°$ to $5°$ of the visual field. The number of faces in the images was varied between 1-6, with a mean of 1.1 ± 0.48 (s.d.). 441 images (1024×768 pixels) were used in these experiments altogether. Of these, 291 images were unique. The remaining 150 stimuli consisted of 50 different images that were repeated twice, but treated uniquely as they were recorded under different experimental conditions. Of the unique images, some were very similar to each other, as only foreground objects but not the background was changed. Since we only counted finding the exact same instance (*i.e.* 1 out of 441) as correct

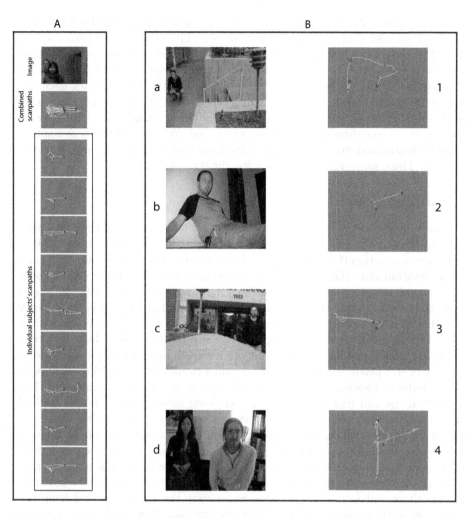

Fig. 1. Examples of scanpaths/stimuli used in the experiment. **A.** Scanpaths of the 9 individual subjects used in the analysis for a given image. The combined fixations of all subjects was used for further analysis of the agreement across all subjects, and for analysis of the ideal subjects' pool size for decoding. The red triangle marks the first and the red square the last fixation, the yellow line the scanpath, and the red circles the subsequent fixations. Top: the image viewed by subjects to generate these scanpaths. The trend of visiting the faces – a highly attractive feature – yields greater decoding performance. **B.** Four example images from the dataset (left) and their corresponding scanpaths for different arbitrary chosen individuals (right). Order is shuffled. See if you can match ("decode") the scanpath to its corresponding images. *The correct answers are: a3, b4, c2 and d1.*

prediction, in at least $\frac{150}{441} \times \frac{2}{440} = 0.15\%$ of cases a nearly correct prediction (same or very similar image) would be counted as incorrect. Hence, our datasets are challenging and the estimates of correct prediction conservative.

Eye-position data were acquired at 1000 Hz using an Eyelink1000 (SR Research, Osgoode, Canada) eye-tracking device. The images were presented on a CRT2 screen (120 Hz), using MATLAB's Psychophysics and eyelink toolbox extensions. Stimulus luminance was linear in pixel values. The distance between the screen and the subject was 80 cm, giving a total visual angle for each image of $28° \times 21°$. Subjects used a chin-rest to stabilize their head. Data were acquired from the right eye alone. Data from a total of nine subjects, each with normal or corrected-to-normal vision, were used. We discard the first fixation from each scanpath to avoid adding trivial information from the initial center fixation. Thus, we worked with $441 \times 9 = 3969$ total scanpaths.

2.2 Decoding Metric

For each image, we created six different "feature maps". Four of the maps were generated using the Itti and Koch saliency map model [8]: (1) combined color-intensity-orientation (CIO) map, (2) color alone (C), (3) intensity alone (I), and (4) orientation alone (O). A "faces" map was generated using the Viola and Jones face recognition algorithm [18]. The sixth map, which we call "CIO+F" was a combination of the face map and the CIO map from the Itti and Koch saliency model, which has been shown to be more predictive of observers fixations than CIO [17]. Each feature map was represented as a positive valued heat map over the image plane, and downsampled substantially, in line with [8], in our case to nine by twelve pixels, each pixel corresponding to roughly 2×2 degrees of visual angle. Subject fixation data was binned into an array of the same size. The saliency maps and fixation data were compared using an ROC-based method [16]. This method compares saliency at fixated and non-fixated locations (see Fig. 2 for an illustration of the method). We assume some threshold saliency level above which locations on the saliency map are considered to be predictions of fixation. If there is a fixation at such a location, we consider it a hit, or true positive. If there is no fixation, it is considered a false positive. We record the true positive and false positive rates as we vary the threshold level from the minimum to the maximum value of the saliency map. Plotting false positive vs. true positive results in a Receiver Operator Characteristics ("ROC") curve. We integrate the Area Under this ROC Curve ("AUC") to get a scalar similiarity measure (AUC of 1 indicates all fixations fall on salient locations, and AUC of 0.5 is chance level). The AUC for the correct scanpath-image pair was ranked against other scanpath-image pairs (from 1 to 31 decoy images, chosen randomly from the remaining 440 to 410 images), and the decoding was considered successful only if the correct image was ranked one. In the largest image set size we tried, if any of the other 31 AUCs for scanpath/images was higher than the one of the correct match, we considered the prediction a miss (e.g. for one decoding trial the algorithm would be as follows: *1.* Randomly select a scanpath out of the 3969 scanpaths. *2.* Consider the image it belongs to, together with 1 to 31 randomly selected decoys. We will attempt to match the scanpath to its associated image out of this set of candidates. *3.* Compute a feature map for each image in the candidate set. *4.* Compute the AUC of the scanpath for each of the 2-32 saliency

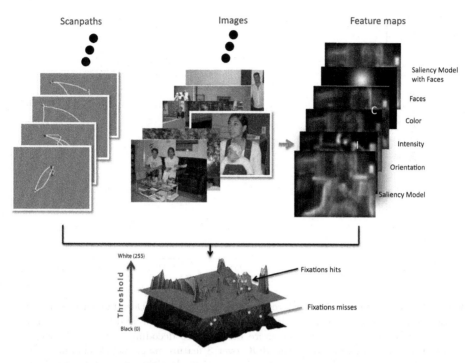

Fig. 2. Illustration of the AUC calculation. For each scanpath, we choose the corresponding image and 1–31 decoys. For each image we calculate each of the 6 feature maps (C, I, O, F, CIO, CIO+F). For a given scanpath and a feature map we then calculate the ROC by varying a threshold over the feature plane and counting how many fixations fall above/below the threshold. The area under the ROC curve (AUC) serves as a measure of agreement between the scanpath and the feature map. We then rank the images by their AUC scores, and consider the decoding correct if the highest AUC is that of the correct image.

maps. 5. Decoding is considered successful iff the image on which the scanpath was actually recorded has the highest AUC score.).

3   Results

We calculated the average success rate of prediction trials, each of which consists of (1) fixations pooled over 9 subjects' scanpaths, and (2) an image set of particular cardinality, from 2 to 32, ranked according to the ROC-fixation score on one of three possible feature maps: CIO, CIO+F, or F. We used the face channel although it carries some false identifications of faces, and some misses, as it has been shown to have higher predictive power, involving high-level (semantic) saliency content with bottom-up driven features [17]. We reasoned that using the face channel alone in this discriminability experiment would provide a novel method of comparing it to saliency maps' predictive power.

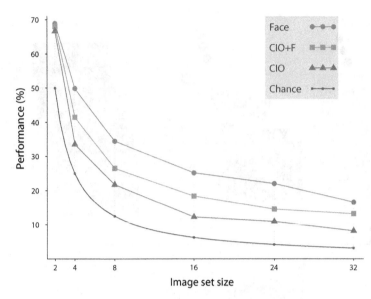

Fig. 3. Decoding performance with respect to image pool size. Decoding with scanpaths pooled over 9 subjects, we varied the number of decoy images used between 1 and 31. The larger the image set size, the more difficult the decoding. For each image set size and scanpath we calculated the ROC over 3 feature maps: a face-channel which is the output of the Viola and Jones face-detection algorithm with the given image (F), a saliency map based on the color, orientation and intensity maps (CIO), and a saliency map combining the face-channel and the color, orientation and intensity maps (CIO+F). While all feature maps yielded a similar decoding performance for the smaller pool size, the performance was least degraded for the F map. The face feature map is higher than the CIO+F map and the two are higher than the CIO map. All maps predict above chance level – shown in the bottom line as the multiplicative inverse of the image set size.

For one decoy per image set (image set size = two), we find that the face feature map (F) was used to correctly predict the image seen by the subjects in 69% of the trials ($p < 10^{-15}$, sign test[1]), while the CIO+F feature map was correct in 68% ($p < 10^{-14}$), and CIO in 66% ($p < 10^{-12}$) of trials. This $F > CIO + F > CIO$ trend persists through all image set sizes. Pooling prediction trials over all image set sizes (6 sizes × 441 trials per size = 2646 trials), we find that using the F map yields a prediction that is at least as accurate as the CIO map in 89.9% of trials, with significance $p < 10^{-8}$ using the sign-test. Similarly, F is at least as predictive as CIO+F in 90.3% of trials ($p < 10^{-15}$), and CIO+F is at least as predictive as CIO in 97.8% of trials ($p < 10^{-21}$). All data points

---

[1] The sign-test tests against the null hypothesis that the distribution of correct decodings is drawn from a binary distribution (50% for the choice of 1 of 2 images, 33% in the case of 1 of 3 images, and so forth up to 3% in the case of 1 out of 32 images). This is the most conservative estimate; additional assumptions on the distribution would yield lower p-values.

in Fig. 3 are significantly above their corresponding chance levels, with the least significant point corresponding to predictions using CIO with image set size 4: this results in correct decoding in 33.6% of trials, compared to 25% for chance, with null hypothesis that predictions are 25% correct being rejected at $p < 10^{-4}$.

We also tested the prediction rates when fixations were pooled over progressively fewer subjects, instead of only nine as above. For this, we used only the CIO+F map (although the face channel shows the highest decoding performance we wanted to use a feature map that combines bottom-up features to match common attention prediction methods), and binary image trials (one decoy). One might imagine that pooling over fixation recordings from different subjects

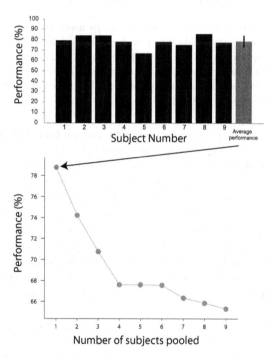

Fig. 4. Performance of the 9 individual subjects. **Upper panel.** For the 441 scanpaths/images, we computed the decoding performance of each individual subject. Bars indicate the performance of each subject. Red bar on the right indicates the average performance of all 9 subjects, with standard error bar. Average subject performance was 79%, with the lowest decoding performance at 67% (subject 4), and the highest at 86% (subject 8). All values are significantly above chance (50%), with p values (sign test) below 10^{-10}. **Lower panel.** Performance of various combinations of the 9 subjects. Scanpaths of 1, 2, ... 9 subjects used to determine the performance differences by using average scanpaths of multiple subjects. The performance of individual subjects shown on the leftmost point is the average of each subjects' performance as shown in the upper panel. The rightmost point is the performance of all subjects combined. Each subject pool was combined from a random choice of subjects out of the 9, reaching the pool size.

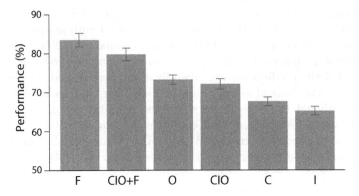

Fig. 5. Decoding performance based on feature maps used. We show the average decoding performance on binary trials using each of the 6 different feature maps, and in each trial, the scanpath of only one individual subject. Thus, for instance, the performance of the CIO+F map is exactly that shown in the average bar in Fig. 4. The higher the performance the more useful the feature is in the decoding. The face channel is the most important one for this dataset.

would increase the signal to noise ratio, but in fact we find that prediction performance only decreases (Fig. 4) with more subjects. There are several possible explanations for this decrease. First, in computing the AUC, we record a correct detection ("hit") whenever a superthreshold saliency map cell overlaps with at least one fixation, but discard information about multiple fixations at that location (*i.e.*, a cell is either occupied by a fixation or not). Thus, the accuracy of the ROC AUC agreement between a saliency map and the fixations of multiple observers degrades with overlapping fixations. As the number of overlapping fixations increases with observers, the reliability of our decoding measure decreases. Indeed, other measures taking into account this phenomenon then can outperform the present metric. Second, if different observers exhibit distinct feature preferences (say, some prefer "color", some prefer "orientation", etc.), the variability in the locations of such features across an image set would contribute to the prediction in this set. It is possible that an image set is more varied along the preferences of any one observer on average than along the pooled preferences of multiple observers. This would make it more difficult to decode from aggregate fixation sets.

The mean percentage of correct decoding for a single subject was 79% (chance is 50%), ($p < 10^{-288}$, sign test). For all combinations of 1 to 9 subjects used, the prediction was above chance (with p values below $p < 10^{-10}$). The lowest prediction performance results from pooling over all nine subjects, with 66% hit rate (still significantly above chance at 50%). Figure 4 shows the prediction for each of the 9 subjects with the CIO+F feature map.

Finally, in order to test the relative contribution of each feature map to the decoding, we used our new decoding correctness rate to compare feature map types, from most discriminating to least. This was done by comparing separately each of the 6 features maps' average decoding performance for binary trials with 9 individual subjects' scanpaths. The results (Fig. 5) show that out of the 6 feature maps the face channel has the highest performance (decoding performance of 82%, $p = 0$) (as shown also in Fig. 3), and the intensity map has the lowest performance (decoding performance: 65%, $p < 10^{-104}$, sign test). All values are significantly above chance (50%).

4   Discussion

In this study, we investigated if scanpath data could be used to decode which image an observer was viewing given only the scanpath and saliency maps. The results were quite strong: in an experiment with 441 trials, each consisting of 32 images with scanpath data belonging to one unknown image in the set, in 73 trials (17%) the correct image was selected, a fraction much higher than chance ($\frac{1}{32} = 3\%$). This leads us to propose a new metric for quantifying the efficacy of saliency maps based on image discriminability. For decoding we used the standard area under ROC curve measure with the fixations from 1 to 9 subjects on a feature map generated by popular models for fixations and attention predictions.

The "decodability" of a dataset is a score given to the combined scanpath/stimuli data for a given feature and as such can be used in various ways: we here used the decodability in order to compare ideal combined subjects' scanpath pool and feature maps' predictive power. Furthermore, we can imagine the same method being used to cluster subjects according to features that pertain specifically to them for a given dataset (*i.e.* if a particular set of subjects tends to look more often on an area in the images than other [19], or tends to fixate on a certain object/target more [20,21,22], this would result in a higher decoding performance for that feature map), or as a measure of the relative amount of stimuli needed to reach a certain level of decoding performance. Our data suggests that clustering by such features to segregate between autistic and normal subjects is perhaps possible based on differences in their looking at faces/objects [21]. However, our autism subjects fixations dataset is too small to reach significance.

In line with earlier results, ours show that saliency maps using bottom-up features such as color, orientation, and intensity are relatively accurate predictors of fixation [16,23,24,25,26] with a performance above 70% (Fig. 5, similar to the estimate in [15]). Adding the information from a face detector boosts performance to over 80%, similar to the estimate in [17]. It is possible that incorporating more complex, higher-level feature maps [27,28] could further improve performance.

Some of the images we used were very similar to each other, and so the image set could be considered challenging. Using this novel decoding metric on larger, more diverse datasets could yield more striking distinctions between the feature maps and their relative contributions to attentional allocation.

Notice that in the results, in particular in Fig. 3, we computed average predictive performance using fixations pooled over all 9 scanpaths recorded per image. However, as we have shown that individual subjects' fixations are more predictive due to variability issues, these results should be even stronger than those we have included above.

A possibility for subsequent work is the prediction not of particular images from a set, but of image content. For example, is it possible to predict whether or not an image contains a face, text, or other specific semantic content based only on the scanpaths of subjects? The same kinds of stereotypical patterns we used to predict images would be useful in this kind of experiment.

Finally, one can think of more sophisticated algorithms for predicting scan-path/image pairs. For instance, one could use information about previously decoded images for future iterations (perhaps by eliminating already decoded images from the pool, making harder decoding more feasible), or a softer rank-ing algorithm (here we considered decoding correct only if the corresponding scanpath was ranked the highest among 32 images; one could, however, com-pute statistics from a soft "confusion matrix" containing all rankings so as to reduce the noise from spuriously high similarity pairs).

We demonstrated a novel method for estimating the similarity between a given set of scanpaths and images by measuring how well scanpaths could de-code the images that corresponded to them. Our decoder ranked images accord-ing to saliency map/fixation similarity, yielding the most similar image as its prediction. While our decoder already yields high performance, there are more sophisticated distance measures that might be more accurate, such as ones used in electrophysiology [7].

Rating a saliency map relative to a scanpath based on its usability as a de-coder for the input stimulus represents a robust new measure of saliency map efficacy, as it incorporates information about how dissimilar a map is from those computed on other images. This novel method can also be used for assessing images sets, for measuring the performance and attention allocation for a given set, for comparing existing saliency map performance measures, and as a metric for the evaluation of eye-tracking data against other psychophysical data.

Acknowledgements

This research was funded by the Mathers Foundation, NGA and NIMH.

References

1. Young, M., Yamane, S.: Sparse population coding of faces in the inferotemporal cortex. Science 256(5061), 1327–1331 (1992)
2. Schwartz, E., Desimone, R., Albright, T., Gross, C.: Shape Recognition and Inferior Temporal Neurons. Proceedings of the National Academy of Sciences of the United States of America 80(18), 5776–5778 (1983)
3. Sato, T., Kawamura, T., Iwai, E.: Responsiveness of inferotemporal single units to visual pattern stimuli in monkeys performing discrimination. Experimental Brain Research 38(3), 313–319 (1980)
4. Perrett, D., Rolls, E., Caan, W.: Visual neurones responsive to faces in the monkey temporal cortex. Experimental Brain Research 47(3), 329–342 (1982)

5. Logothetis, N., Pauls, J., Poggio, T.: Shape representation in the inferior temporal cortex of monkeys. Current Biology 5(5), 552–563 (1995)
6. Hung, C., Kreiman, G., Poggio, T., DiCarlo, J.: Fast Readout of Object Identity from Macaque Inferior Temporal Cortex (2005)
7. Quiroga, R., Reddy, L., Koch, C., Fried, I.: Decoding Visual Inputs From Multiple Neurons in the Human Temporal Lobe. Journal of Neurophysiology 98(4), 1997 (2007)
8. Itti, L., Koch, C., Niebur, E., et al.: A model of saliency-based visual attention for rapid scene analysis. IEEE Transactions on Pattern Analysis and Machine Intelligence 20(11), 1254–1259 (1998)
9. Dickinson, S., Christensen, H., Tsotsos, J., Olofsson, G.: Active object recognition integrating attention and viewpoint control. Computer Vision and Image Understanding 67(3), 239–260 (1997)
10. Koch, C., Ullman, S.: Shifts in selective visual attention: towards the underlying neural circuitry. Hum. Neurobiol. 4(4), 219–227 (1985)
11. Yarbus, A.: Eye Movements and Vision. Plenum Press, New York (1967)
12. Goldstein, R., Woods, R., Peli, E.: Where people look when watching movies: Do all viewers look at the same place? Computers in Biology and Medicine 37(7), 957–964 (2007)
13. Privitera, C., Stark, L.: Evaluating image processing algorithms that predict regions of interest. Pattern Recognition Letters 19(11), 1037–1043 (1998)
14. Itti, L., Koch, C.: Computational modeling of visual attention. Nature Rev. Neurosci. 2(3), 194–203 (2001)
15. Peters, R., Iyer, A., Itti, L., Koch, C.: Components of bottom-up gaze allocation in natural images. Vision Res. 45(18), 2397–2416 (2005)
16. Tatler, B., Baddeley, R., Gilchrist, I.: Visual correlates of fixation selection: effects of scale and time. Vision Research 45(5), 643–659 (2005)
17. Cerf, M., Harel, J., Einhäuser, W., Koch, C.: Predicting human gaze using low-level saliency combined with face detection. In: Platt, J., Koller, D., Singer, Y., Roweis, S. (eds.) Advances in Neural Information Processing Systems, vol. 20. MIT Press, Cambridge (2008)
18. Viola, P., Jones, M.: Rapid object detection using a boosted cascade of simple features. Computer Vision and Pattern Recognition 1, 511–518 (2001)
19. Buswell, G.: How People Look at Pictures: A Study of the Psychology of Perception in Art. The University of Chicago press (1935)
20. Barton, J.: Disorders of face perception and recognition. Neurol. Clin. 21(2), 521–548 (2003)
21. Klin, A., Jones, W., Schultz, R., Volkmar, F., Cohen, D.: Visual Fixation Patterns During Viewing of Naturalistic Social Situations as Predictors of Social Competence in Individuals With Autism (2002)
22. Adolphs, R.: Neural systems for recognizing emotion. Curr. Op. Neurobiol. 12(2), 169–177 (2002)
23. Baddeley, R., Tatler, B.: High frequency edges (but not contrast) predict where we fixate: A Bayesian system identification analysis. Vision Research 46(18), 2824–2833 (2006)
24. Einhäuser, W., König, P.: Does luminance-contrast contribute to a saliency map for overt visual attention?. Eur. J. Neurosci. 17(5), 1089–1097 (2003)
25. Einhäuser, W., Kruse, W., Hoffmann, K., König, P.: Differences of monkey and human overt attention under natural conditions. Vision Res. 46(8-9), 1194–1209 (2006)

26. Navalpakkam, V., Itti, L.: Search goal tunes visual features optimally. Neuron 53(4), 605–617 (2007)
27. Kayser, C., Nielsen, K., Logothetis, N.: Fixations in natural scenes: Interaction of image structure and image content. Vision Res. 46(16), 2535–2545 (2006)
28. Einhäuser, W., Rutishauser, U., Frady, E., Nadler, S., König, P., Koch, C.: The relation of phase noise and luminance contrast to overt attention in complex visual stimuli. J. Vis. 6(11), 1148–1158 (2006)

A Novel Hierarchical Framework for Object-Based Visual Attention

Rebecca Marfil, Antonio Bandera, Juan Antonio Rodríguez,
and Francisco Sandoval

Departamento de Tecnología Electrónica,
E.T.S.I. Telecomunicación, Universidad de Málaga
Campus de Teatinos, 29071-Málaga, Spain
rebeca@uma.es

Abstract. This paper proposes an artificial visual attention model which builds a saliency map associated to the sensed scene using a novel perception-based grouping process. This grouping mechanism is performed by a hierarchical irregular structure, and it takes into account colour contrast, edge and depth information. The resulting saliency map is composed by different parts or 'pre-attentive objects' which correspond to units of visual information that can be bound into a coherent and stable object. Besides, the ability to handle dynamic scenarios is included in the proposed model by introducing a tracking mechanism of moving objects, which is also performed using the same hierarchical structure. This allows to conduct the whole attention mechanism in the same structure, reducing the computational time. Experimental results show that the performance of the proposed model is compatible with the existing models of visual attention whereas the object-based nature of the proposed approach renders advantages of precise localization of the focus of attention and proper representation of the shapes of the attended 'pre-attentive objects'.

1 Introduction

In biological vision systems, the attention mechanism is responsible of selecting the relevant information from the sensed field of view so that the complete scene can be analyzed using a sequence of rapid eye saccades [1]. In the recent years, efforts have been made to imitate such attention behavior in artificial vision systems, because it allows to optimize the computational resources as they can be focused on the processing of a set of selected regions only. Probably one of the most influential theoretical models of visual attention is the spotlight metaphor [2], by which many concrete computational models have been inspired [3][4][5]. These approaches are related with the *feature integration theory*, a biologically plausible theory proposed to explain human visual search strategies [6]. According to this model, these methods are organized into two main stages. First, in a preattentive task-independent stage, a number of parallel channels compute image features. The extracted features are integrated into a single saliency map which codes the saliency of each image region. The most salient regions are selected from this map. Second, in an attentive task-dependent stage, the spotlight

L. Paletta and J.K. Tsotsos (Eds.): WAPCV 2008, LNAI 5395, pp. 27–40, 2009.

is moved to each salient region to analyze it in a sequential process. Analyzed regions are included in an *inhibition map* to avoid movement of the spotlight to an already visited region. Thus, while the second stage must be redefined for different systems, the preattentive stage is general for any application. Although these models have good performance in static environments, they cannot in principle handle dynamic environments due to their impossibility to take into account the motion and the occlusions of the objects in the scene. In order to solve this problem, an attention control mechanism must integrate depth and motion information to be able to track moving objects. Thus, Maki et al. [7] propose an attention mechanism which incorporates depth and motion as features for the computation of saliency and Itti [8] incorporates motion and flicker channels in its model.

The previously described methods deploy attention at the level of space locations (*space-based models of visual attention*). The models of space-based attention scan the scene by shifting attention from one location to the next to limit the processing to a variable size of space in the visual field. Therefore, they have some intrinsic disadvantages. In a normal scene, objects may overlap or share some common properties. Then, attention may need to work in several discontinuous spatial regions at the same time. If different visual features, which constitute the same object, come from the same region of space, an attention shift will be not required [9]. On the contrary, other approaches deploy attention at the level of objects. *Object-based models of visual attention* provide a more efficient visual search than space-based attention. Besides, it is less likely to select an empty location. In the last few years, these models of visual attention have received an increasing interest in computational neuroscience and in computer vision. Object-based attention theories are based on the assumption that attention must be directed to an object or group of objects, instead to a generic region of the space [10]. Therefore, these models will reflect the fact that the perception abilities must be optimized to interact with objects and not just with disembodied spatial locations. Thus, visual systems will segment complex scenes into objects which can be subsequently used for recognition and action.

Finally, space-based and object-based approaches are not mutually exclusive, and several researchers have proposed attentional models that integrate both approaches. Thus, in the Sun and Fisher's proposal [9], the model of visual attention combines object- and feature-based theories. In its current form, this model is able to replicate human viewing behaviour. However, it needs that input images will be manually segmented. That is, it uses information that is not available in a preattentive stage, before objects are recognized [10].

This paper presents an object-based model of visual attention, which is capable of handling dynamic environments. The proposed system integrates bottom-up (data-driven) and top-down (model-driven) processing. The bottom-up component determines and selects salient 'pre-attentive objects' by integrating different features into the same hierarchical structure. These 'pre-attentive objects' or 'proto-objects' [11][10] are image entities which do not necessarily correspond with a recognizable object, although they possess some of the characteristics of objects. Thus, it can be considered that they are the result of the initial segmentation of the image input into candidate objects (i.e. grouping together those input pixels which are likely to correspond to parts of the same object in the real world, separately from those which are likely to belong to other objects). This is the main contribution of the proposed approach, as it is able to group the image pixels into entities which can be considered as *segmented perceptual units* [10]. On the other hand, the top-down component could make use of object templates to filter out data and shift the attention to objects which are relevant to accomplish the current tasks to reach. However, it must be noted that this work is mainly centered in the task-independent stage of the model of visual attention. Therefore, the experiments are restricted to bottom-up mode. Finally, in a dynamic scenario, the locations and shapes of the objects may change due to motion and minor illumination differences between consecutive acquired images. In order to deal with these scenes, a tracking approach for 'inhibition of return' is employed in this paper. This approach is conducted using the same hierarchical structure and its application to this framework is the second main novelty of the proposed model.

The remainder of the paper is organized as follows: Section 2 provides an overview of the proposed method. Section 3 presents a description of the computation of the saliency map using a hierarchical grouping process. The proposed mechanism to implement the inhibition of return is described in Section 4. Section 5 deals with some obtained experimental results. Finally, conclusions and future works are presented in Section 6.

2   Overview of the Proposed Model

Fig. 1 shows an overview of the proposed architecture. The visual attention model we propose employs a concept of salience based on 'pre-attentive objects'. These 'pre-attentive objects' are defined as the blobs of uniform color and disparity of the image which are bounded by the edges obtained using a Canny detector. To obtain these entities, the proposed method has two main stages. In the first stage the input image pixels are grouped into blobs of uniform colour. These regions constitute an efficient image representation that replace the pixel-based image representation. Besides, these regions preserve the image geometric structure as each significant feature contain at least one region. In the second stage, this set of blobs is grouped into a smaller set of 'pre-attentive objects' taking into account not only the internal visual coherence of the obtained blobs but also the external relationships among them. These two stages are accomplished by means of an irregular pyramid: the Bounded Irregular Pyramid (BIP). The BIP combines the 2x2/4 regular structure with an irregular simple graph [13]. In the first stage - called pre-segmentation stage- the proposed approach generates a first set of pyramid levels where nodes are grouped using a colour-based criterion. Then, in the second stage -perceptual grouping stage- new pyramid

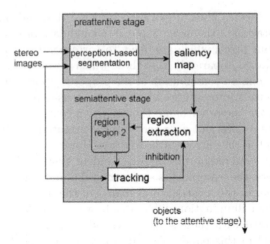

Fig. 1. Overview of the proposed model of visual attention. It has two main stages: a preattentive stage in which the input image pixels are grouped into a set of 'preattentive objects' and a semiattentive stage where the inhibition of return is implemented using a tracking process.

levels are generated over the previously built BIP. The similarity among nodes of these new levels is defined using a more complex distance which takes into account information about their common boundaries and internal features like their colour, size or disparity.

A 'pre-attentive object' catches the attention if it differs from its immediate surrounding. In our model, we compute a measure of bottom-up salience associated to each 'pre-attentive object' as a distance function which takes into account colour and luminosity contrasts between the 'pre-attentive object' and all the objects in its surrounding. Then, the focus of attention is changed depending of the shapes of the objects in the scene. This is more practical that to maintain a constant size of the focus of attention [10].

Finally, the general structure of the model of visual attention is related to a previous proposal of Backer and Mertsching [12]. Thus, although we do not compute parallel features at the preattentive stage, this stage is followed by a semiattentive stage where a tracking process is performed. Besides, while Backer and Mertsching's approach performs the tracking over the saliency map by using dynamics neural fields, our method tracks the most salient regions over the input image using a hierarchical approach based on the Bounded Irregular Pyramid [14]. The output regions of the tracking algorithm are used to implement the 'inhibition of return' which avoids revisiting recently attended objects. The main disadvantage of using dynamic neural fields for controlling behavior is the high computational cost for simulating the field dynamics by numerical methods. The Bounded Irregular Pyramid approach allows fast tracking of a non-rigid object without a previous learning of different objects views [14].

3   Estimation of the Saliency Map

Most of the models of visual attention build different scales of the input image and determine saliency by considering the vicinity of the individual pixels in these scales. As it has been pointed out by Aziz and Mertsching [1], the use of such coarse-to-fine scales during feature extraction provokes fuzziness in the final conspicuity map. This drawback can be avoided by adopting a region-based methodology for the model of attention. These region-based approaches usually work at a single scale, on blobs defined in the input image. As it has been previously mentioned, the pixels of each blob are grouped together to define 'pre-attentive objects' in the scene. Following this guideline, the proposed model of attention segments the image into perceptually uniform blobs first and then it computes features on these 'pre-attentive objects'. Therefore, a segmentation step is a pre-requisite for our model of visual attention. However, contrary to other approaches [1][10], the proposed approach associates a hierarchical representation to each detected 'pre-attentive object'. In this paper, this representation will be very useful to implement a fast mechanism of 'inhibition of return'. Future extensions of this work will include a fast template matching algorithm which will make also use of these hierarchical representations to enhance the salience of 'pre-attentive objects' with a shape similar to the objects which are relevant to accomplish the current tasks to execute (top-down component of the model of visual attention).

3.1   Pre-segmentation Stage

Pyramids are hierarchical structures which have been widely used in segmentation tasks [15]. Instead of performing image segmentation based on a single representation of the input image, a pyramid segmentation algorithm describes the contents of the image using multiple representations with decreasing resolution. Pyramid segmentation algorithms exhibit interesting properties when compared to segmentation algorithms based on a single representation. Thus, local operations can adapt the pyramid hierarchy to the topology of the image, allowing the detection of global features of interest and representing them at low resolution levels [16].

The Bounded Irregular Pyramid (BIP) [13] is a mixture of regular and irregular pyramids whose goal is to combine their advantages. A 2x2/4 regular structure is used in the homogeneous regions of the input image and a simple graph structure in the non-homogeneous ones. The mixture of both structures generates an irregular configuration which is described as a graph hierarchy in which each level $G_l = (N_l, E_l)$ consists of a set of nodes, N_l, linked by a set of intra-level edges E_l. Each graph G_{l+1} is built from G_l by computing the nodes of N_{l+1} from the nodes of N_l and establishing the inter-level edges $E_{l,l+1}$. Therefore, each node n_i of G_{l+1} has associated a set of nodes of G_l, which is called the *reduction window* of n_i. This includes all nodes linked to n_i by an inter-level edge. The node n_i is called *parent* of the nodes in its reduction window, which are called *sons*. The successive levels of the hierarchy are built using a regular decimation process and a union-find strategy. Therefore, there are two types of nodes: nodes belonging to the 2x2/4 structure, named regular nodes, and virtual nodes or nodes belonging to the irregular structure. In any case, two nodes $n_i \in N_l$ and $n_j \in N_l$ which are neighbors at level l are linked by an intra-level edge $e_{ij} \in E_l$.

The proposed approach uses a BIP structure to accomplish the detection of the 'pre-attentive objects' and the subsequent computation of the saliency map following a perceptual grouping approach. In this hierarchy, the first levels perform the pre-segmentation stage using a colour-based distance to group pixels into homogeneous blobs. In order to introduce colour information within the BIP, all the nodes of the structure have associated 2 parameters: chromatic phasor $S_{\angle H}(n)$, and luminosity $V(n)$, where S, H and V are the saturation, hue and value of the HSV colour space. The chromatic phasor and the luminosity of a node n at level l are equal to the average of the chromatic phasors and luminosity values of the nodes in its reduction window, i.e. the nodes of the level l-1 which are linked to n.

The employed similarity measurement between two nodes is the HSV colour distance. Thus, two nodes are similar or have a similar colour if the distance between their HSV values is less than a similarity threshold T.

The graph $G_0 = (N_0, E_0)$ is a 8-connected graph where the nodes are the pixels of the original image. The chromatic phasors and the luminosity values of the nodes in $G_0 = (N_0, E_0)$ are equal to the chromatic phasors and luminosity values of their corresponding image pixels. The process to build the graph $G_{l+1} = (N_{l+1}, E_{l+1})$ from $G_l = (N_l, E_l)$ is briefly described below (see [13] for further details):

1. *Regular decimation process.* If four regular neighbor nodes of the level l have similar colour, they are grouping together, generating a regular node in $l+1$.
2. *Parent search and intra-level twining.* Once the regular structure is generated, there are some regular orphan nodes (regular nodes without parent). From each of these nodes (i, j, l), a search is made for the most similar node with parent in its neighborhood $\xi_{(i,j,l)}$. If this neighbor node is found, the node (i, j, l) is linked to the parent of this neighbor node. On the contrary, if for this node a parent is not found, then a search is made for the most similar neighbor node without parent to link to it. If this node is found, then both nodes are linked, generating a virtual node at level $l + 1$.
3. *Virtual parent search and virtual node linking.* Each virtual orphan node n_i searches for the most similar node with parent in its neighborhood ξ_{n_i}. If for n_i a parent is found, then it is linked to it. On the other hand, if a parent is not found, the virtual orphan node n_i looks for the most similar orphan node in its neighborhood to generate a new virtual node at level $l + 1$. The only restriction to this step is that the parent of a virtual node must be always a virtual node so a virtual node cannot be linked to a regular parent. It allows to preserve the regular nature of the regular part of the BIP.

The hierarchy stops to grow when it is no longer possible to link together any more nodes because they are not similar. In order to perform the

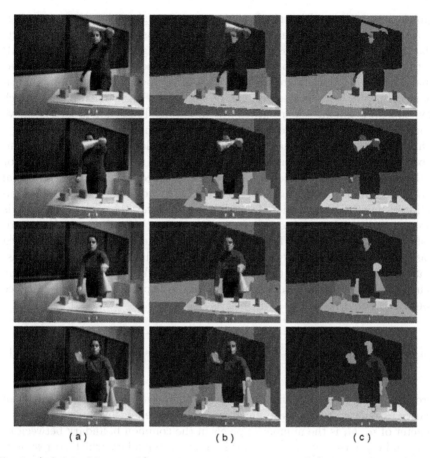

Fig. 2. a) Original images; b) pre-segmentation images; and c) obtained regions after the perceptual grouping

pre-segmentation, the orphan nodes are used as roots. The described method has been tested and compared with other similar pyramid approaches for colour image segmentation [15]. This comparative study concludes that the BIP runs faster than other irregular approaches when benchmarking is performed in a standard sequential computer. Besides, the BIP obtains similar results than the main irregular structures. Fig. 2.b shows the pre-segmentation images associated to the images in Fig. 2.a. It can be noted as the input images are correctly segmented into blobs of uniform colour.

3.2   Perceptual Grouping Stage

After the local similarity pre-segmentation stage, grouping blobs aims at simplifying the content of the obtained partition in order to obtain the set of final 'preattentive objects'. Two constraints are taken into account for an efficient grouping process: first, although all groupings are tested, only the best groupings are locally retained; and second, all the groupings must be spread on the image so that no part of the image is advantaged. For managing this process, the BIP structure is employed: the roots of the pre-segmented blobs are considered as virtual nodes which constitute the first level of the perceptual grouping multi-resolution output. Successive levels can be built using the virtual parent search and virtual node linking scheme previously described. Finally, in order to achieve the perceptual grouping process, a specific distance must be defined.

This distance has three main components: the colour contrast between image blobs, the edges of the original image computed using the Canny detector and the disparity of the image blobs. In order to speed up the process, a global contrast measure is used instead of a local one. It avoids working at pixel resolution, increasing the computational speed. In this distance the contrast measure is complemented with internal regions properties and with attributes of the boundary shared by both regions. To perform correctly, the nodes of the BIP which are associated to the perceptual grouping multi-resolution output store statistics about the HSV values of the roots generated by the pre-segmentation stage which are linked to them and about their mean disparity. Then, the distance between two nodes n_i and n_j is defined as

$$\Upsilon(n_i, n_j) = \sqrt{w_1(\cdot\frac{d(n_i, n_j) \cdot b_i}{\alpha \cdot (c_{ij}) + (\beta \cdot (b_{ij} - c_{ij}))})^2 + w_2(\cdot disp(n_i) - disp(n_j))^2} \quad (1)$$

where $d(n_i, n_j)$ is the colour distance between n_i and n_j and $disp(x)$ is the mean disparity associated to the base image region represented by node x. b_i is the perimeter of n_i, b_{ij} is the number of pixels in the common boundary between n_i and n_j and c_{ij} is the set of pixels in this common boundary which corresponds to pixels of the boundary detected with the Canny detector. α and β are to constants values used to control the influence of the Canny edges in the grouping process. We set these parameters to 0.1 and 1.0 respectively. In the same way w_1 and w_2 are two constant values which weight the terms associated with the colour and the disparity. In our case they are set to 0.5 and 1.0, respectively.

In order to build a new hierarchy level G_{l+1}, the virtual parent search and virtual node linking process described in Section 3.1 is applied. However, a different threshold value T_{perc} is employed. The grouping process is iterated until the number of nodes remains constant between two successive levels.

After the pre-segmentation and perceptual grouping stages, the nodes of the BIP with no parent will be the roots of the 'pre-attentive objects'. It must be appreciated that these 'pre-attentive objects' can be represented as hierarchical structures, where the object root constitutes the higher level of the representation and the nodes of the input image linked to this root conform its lower level. Thus, Fig. 3 presents an example of hierarchical representation of a 'pre-attentive object' (only regular nodes are displayed). The base of the pyramid (level 0) contains 64x64 pixels. Fig. 3 shows how pixels at level l are arranged into sets of 2x2 elements to create a regular node at level $l+1$.

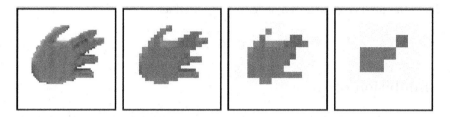

Fig. 3. 'Pre-attentive object' hierarchical representation. Each image represents a consecutive level of the regular part of the BIP representation of a hand, starting at level 0.

Finally, Fig. 2.c shows the set of 'pre-attentive objects' associated to the images in Fig. 2.a. It can be noted that the obtained regions do not always correspond to the set of natural objects presented in the image, but they provide an image segmentation which is more coherent with the human-based image decomposition.

3.3 Computation of the Saliency Values

Once the input image is divided into a set of perceptually uniform blobs or 'pre-attentive objects', we compute the saliency of each blob using a colour contrast and an intensity contrast measures. As each of these blobs corresponds with a root of the BIP structure previously generated, it contains all the necessary information about the concerned image region such as its average chromatic phasor and intensity and the set of neighbors of this blob.

Then, we compute the colour contrast of a 'pre-attentive object' $\mathcal{R}_i$ as the mean colour gradient MCG_i along its boundary to the neighbor blobs:

$$MCG_i = \frac{S_i}{b_i} \sum_{j \in N_i} b_{ij} * d(<C_i>, <C_j>) \qquad (2)$$

being b_i the perimeter of $\mathcal{R}_i$, N_i the set of regions which are neighbors of $\mathcal{R}_i$, b_{ij} the length of the perimeter of the region $\mathcal{R}_i$ in contact with the region $\mathcal{R}_j$, $d(<C_i>, <C_j>)$ the colour distance between the colour mean values $<C>$ of the regions $\mathcal{R}_i$ and $\mathcal{R}_j$ and S_i the mean saturation value of the region $\mathcal{R}_i$.

The use of S_i in the MCG avoids that colour regions with low saturation (grey regions) obtain a higher value of colour contrast than pure colour regions. The problem is that white, black and pure grey regions are totally suppressed. To take into account these regions, the luminosity contrast is computed. The luminosity contrast of a region $\mathcal{R}_i$ is the mean luminosity gradient MIG_i along its boundary to the neighbor regions:

$$MLG_i = \frac{1}{b_i} \sum_{j \in N_i} b_{ij} * d(<I_i>, <I_j>) \qquad (3)$$

being $<I_i>$ the mean luminosity value of the region $\mathcal{R}_i$.

Then the final color salience of $\mathcal{R}_i$ is computed as:

$$MG_i = \sqrt{MCG_i^2 + MLG_i^2} \tag{4}$$

4   Inhibition of Return

Human visual psychophysics studies have demonstrated that a local inhibition is activated in the saliency map to avoid attention being directed immediately to a previously attended region. This mechanism is usually called the 'inhibition of return', and, in the context of artificial models of visual attention, it has been usually implemented using a 2D inhibition map that contains suppression factors for one or more focuses of attention that were recently attended [17][18]. However, this 2D inhibition map is not able to handle the situations where inhibited objects are in motion or when the vision system itself is in motion. Dynamic scenes require to be handled with a totally different process in comparison to static scenes because the locations and shapes of the objects may change due to motion and minor illumination differences between consecutive frames. In this situation, establishing a correspondence between regions of the previous frame with those of the successive frame becomes a significant issue.

In order to allow that the inhibition can track an object while it changes its location, the model proposed by Backer et al. [19] relates the inhibitions to features of activity clusters. However, the scope of dynamic inhibition becomes very limited as it is related to activity clusters rather than objects themselves [1]. Thus, it is a better option to attach the inhibition to moving objects [20]. For instance, the recent proposal of Aziz and Mertsching [1] utilizes a queue of inhibited region features to maintain inhibition in dynamic scenes.

Our system implements an object-based 'inhibition of return'. A list of the last attended 'pre-attentive objects' is maintained at the semi-attentive stage of the visual attention model. This list stores information about the colour and last position of the 'pre-attentive object'. It also stores the last hierarchical representation associated to each 'pre-attentive object'. When the vision system moves, the proposed approach keeps track of the 'pre-attentive objects' that it has already visited. The employed tracking algorithm [14] is also based on the Bounded Irregular Pyramid (BIP). This tracking algorithm allows to track non-rigid objects without a previous learning of different object views in real time. To do that, the method uses weighted templates which follow up the viewpoint and appearance changes of the objects to track. The templates and the targets are represented using BIPs. Thus, the generation of the whole set of 'pre-attentive objects' and the tracking of the attended ones to inhibit them are performed into the same hierarchical framework. This will allow to reduce the computation time associated to the whole model of visual attention.

Then, the most salient 'pre-attentive object' $\mathcal{R}_1$ of the saliency map will constitute the image region to attend. When this region is analyzed, a new saliency map is computed. In this map, the region associated to the new estimated location of $\mathcal{R}_1$ is inhibited. If the image is taken from a static scene, the focus of

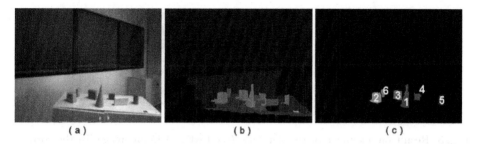

Fig. 4. Saliency map computation and 'pre-attentive objects' selection in a static scene:
a) Left input image; b) saliency map; and c) the five most salient 'pre-attentive objects'

attention will be shifted from one 'pre-attentive object' to another. Thus, Fig. 4.c
shows the 'pre-attentive objects' associated to the saliency map in Fig. 4.b which
will be selected after five focusing steps.

5   Experimental Results

In order to test the performance of the proposed model of visual attention, we
have compared it with the model of Itti et al. [17], using the same database
of images they use [1]. This database consists of 64 colour images which contain
an emergency triangle [21]. Our results show that in 64% of the images a 'pre-
attentive object' inside the emergency triangle is chosen as the most salient
image region (see Fig. 5 for one example).

The proposed model of visual attention has been also examined through video
sequences which include humans and moving objects in the scene. Fig. 6 shows
a sample image sequence seen by a stationary binocular camera head. Every
10th frame is shown. The attended 'pre-attentive object' is marked by a black
and white bounding-box in the input frames. 'Pre-attentive objects' which are
inhibited are marked by a black bounding-box. If the tracking algorithm detects
that a 'pre-attentive object' has suffered a high shape deformation, it will be
not inhibited. Thus, it can be noted that inhibited blobs, like the green cone at
frame 11 or the hand at frame 51, are attended at different sequence frames. It
must be also noted that the activity follows the 'pre-attentive objects' closely,
preventing the related templates from being corrupted by occlusions. Backer
and Mertsching [12] propose to solve the occlusion problem with the inclusion
of depth information. However, depth estimation is normally corrupted by noise
and is often coarsely calculated in order to bound the computational complexity.
In our approach, the tracker is capable of handling scale changes, object defor-
mations, partial occlusions and changes of illumination. Thus, for instance, in
frame 81, an occlusion of the green cone is correctly handled by the tracking
algorithm, which is capable to recover the object before frame 91. Finally, it can

---

[1] http://ilab.usc.edu/imgdbs/

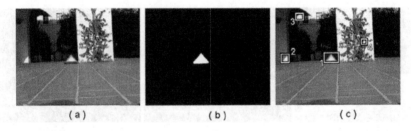

Fig. 5. Result on an image of the database (see text): a) Input image; b) the corresponding binary mask; and c) the four most relevant 'pre-attentive objects'

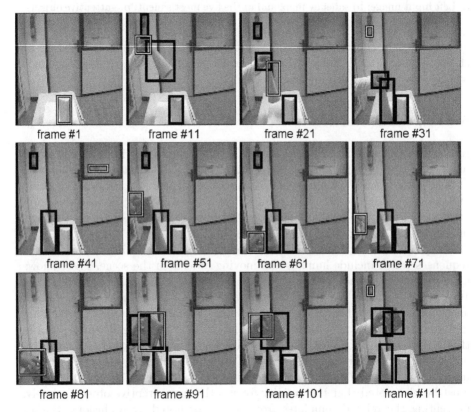

frame #1   frame #11   frame #21   frame #31

frame #41   frame #51   frame #61   frame #71

frame #81   frame #91   frame #101   frame #111

Fig. 6. Left input images of a video sequence. Attended 'pre-attentive objects have been marked by black and white bounding-boxes and inhibited ones have been marked by black bounding-boxes.

be appreciated that the focus of attention is directed at certain frames to uninterested regions of the scene. For instance, this phenomenon occurs at frame 41. These regions cannot usually be correctly tracked and they are removed from the list of 'pre-attentive objects' stored at the semi-attentive stage. Other 'pre-attentive objects', like the fire extinguisher at the background of the image, are not correctly tracked as they are usually divided into several, different blobs.

6   Conclusions and Future Work

This paper has presented a visual attention model that integrates bottom–up and top–down processing. It runs at 10 frames per second using 320x240 images on a standard Pentium personal computer when there are less than five inhibited targets. The proposed model employs two selection stages, providing an additional semi-attentive computation stage where the 'inhibition of return' has been performed. Our model divides the visual scene into perceptually uniform blobs or 'pre-attentive objects'. Thus, the model can direct the attention on candidate, real objects, similarly to the behavior observed in humans. These 'pre-attentive objects' are stored at the semi-attentive stage as hierarchical templates. Currently, this representation is used by the fast tracking algorithm that implements the 'inhibition of return'. Future work will use this representation to conduct the top–down process which will change the bottom–up saliency value associated to each 'pre-attentive object' as a function of the general tasks to reach.

In the future, the integration of this mechanism with an attentive stage that will control the field of attention following several behaviors will allow us to incorporate it in a general active vision system. We have recently incorporated the proposed visual attention model mechanism in two different applications which are being developed. The first application is a visual perception system whose main goal is to help in the learning process of a humanoid robot HOAP-I. The second application is a system to autonomously acquire visual landmarks for mobile robot simultaneous localization and mapping.

Acknowledgments

The authors have been partially supported by the Spanish Ministerio de Educación y Ciencia and by the Junta de Andalucía under projects TIN2005-01359 and P06-TIC-02123, respectively.

References

1. Aziz, M.Z., Mertsching, B.: Color saliency and inhibition using static and dynamic scenes in region based visual attention. In: Paletta, L., Rome, E. (eds.) WAPCV 2007. LNCS (LNAI), vol. 4840, pp. 234–250. Springer, Heidelberg (2007)
2. Eriksen, C.W., Yenh, Y.Y.: Allocation of attention in the visual field. Journal of Experimental Psychology: Human Perception and Performance 11(5), 583–597 (1985)
3. Koch, C., Ullman, S.: Shifts selective visual attention: Towards the underlying neural circuitry. Human Neurobiology 4, 219–227 (1985)
4. Milanese, R.: Detecting salient regions in an image: from biological evidence to computer implementation, PhD Thesis, Univ. of Geneva (1993)

5. Itti, L.: Real-time high-performance attention focusing in outdoors color video streams. In: Proc. SPIE Human Vision and Electronic Imaging (HVEI 2002), pp. 235–243 (2002)

6. Treisman, A.M., Gelade, G.: A feature integration theory of attention. Cognitive Psychology 12(1), 97–136 (1980)

7. Maki, A., Nordlund, P., Eklundh, J.O.: Attentional scene segmentation: integrating depth and motion. Computer Vision and Image Understanding 78(3), 351–373 (2000)

8. Itti, L.: Automatic foveation for video compression using a neurobiological model of visual attention. IEEE Trans. on Image Processing 13(10), 1304–1318 (2004)

9. Sun, Y., Fisher, R.B.: Object-based visual attention for computer vision. Artificial Intelligence 146(1), 77–123 (2003)

10. Orabona, F., Metta, G., Sandini, G.: A proto-object based visual attention model. In: Paletta, L., Rome, E. (eds.) WAPCV 2007. LNCS (LNAI), vol. 4840, pp. 198–215. Springer, Heidelberg (2007)

11. Pylyshyn, Z.W.: Visual indexes, preconceptual objects, and situated vision. Cognition 80(1–2), 127–158 (2001)

12. Backer, G., Mertsching, B.: Two selection stages provide efficient object-based attentional control for dynamic vision. In: Paletta, L. (ed.) International Workshop on Attention and Performance in Computer Vision (WAPCV 2003). Joanneum Research, Graz (2003)

13. Marfil, R., Molina-Tanco, L., Bandera, A., Sandoval, F.: The construction of bounded irregular pyramids with a union-find decimation process. In: Escolano, F., Vento, M. (eds.) GbRPR 2007. LNCS, vol. 4538, pp. 307–318. Springer, Heidelberg (2007)

14. Marfil, R., Molina-Tanco, L., Rodríguez, Sandoval, F.: Real-time object tracking using bounded irregular pyramids. Pattern Recognition Letters 28, 985–1001 (2007)

15. Marfil, R., Molina-Tanco, L., Bandera, A., Rodríguez, J.A., Sandoval, F.: Pyramid segmentation algorithms revisited. Pattern Recognition 39(8), 1430–1451 (2006)

16. Huart, J., Bertolino, P.: Similarity-based and perception-based image segmentation. In: Proc. IEEE Int. Conf. on Image Processing, vol. 3(3), pp. 1148–1151 (2005)

17. Itti, L., Koch, U., Niebur, E.: A model of saliency-based visual attention for rapid scene analysis. IEEE Trans. on Pattern Analysis and Machine Intelligence 20, 1254–1259 (1998)

18. Frintrop, S., Backer, G., Rome, E.: Goal-directed search with a top-down modulated computational attention system. In: Kropatsch, W.G., Sablatnig, R., Hanbury, A. (eds.) DAGM 2005. LNCS, vol. 3663, pp. 117–124. Springer, Heidelberg (2005)

19. Backer, G., Mertsching, B., Bollmann, M.: Data- and model-driven gaze control for an active-vision system. IEEE Trans. on Pattern Analysis and Machine Intelligence 23, 1415–1429 (2001)

20. Tipper, S.P.: Object-centred inhibition of return of visual attention. Quarterly Journal of Experimental Psychology 43, 289–298 (1991)

21. Itti, L., Koch, C.: Feature combination strategies for saliency-based visual attention systems. Journal of Electronic Imaging 10(1), 161–169 (2001)

Where Do We Grasp Objects? – An Experimental Verification of the Selective Attention for Action Model (SAAM)

Christoph Böhme and Dietmar Heinke

School of Psychology, University of Birmingham, Birmingham B15 2TT
{cxb632,d.g.heinke}@bham.ac.uk

Abstract. Classically, visual attention is assumed to be influenced by visual properties of objects, e. g. as assessed in visual search tasks. However, recent experimental evidence suggests that visual attention is also guided by action-related properties of objects ("affordances", [1,2]), e. g. the handle of a cup *affords* grasping the cup; therefore attention is drawn towards the handle (see [3] for example). In a first step towards modelling this interaction between attention and action, we implemented the Selective Attention for Action model (SAAM). The design of SAAM is based on the Selective Attention for Identification model (SAIM [4]). For instance, we also followed a soft-constraint satisfaction approach in a connectionist framework. However, SAAM's selection process is guided by locations within objects suitable for grasping them whereas SAIM selects objects based on their visual properties. In order to implement SAAM's selection mechanism two sets of constraints were implemented. The first set of constraints took into account the anatomy of the hand, e. g. maximal possible distances between fingers. The second set of constraints (geometrical constraints) considered suitable contact points on objects by using simple edge detectors. At first, we demonstrate here that SAAM can successfully mimic human behaviour by comparing simulated contact points with experimental data. Secondly, we show that SAAM simulates affordance-guided attentional behaviour as it successfully generates contact points for only one object in two-object images.

1 Introduction

Actions need to be tightly guided by vision in our daily interactions with our environment. To maintain such a direct guidance, Gibson postulated that the visual system automatically extract "affordances" of objects [2]. According to Gibson, affordance refers to parts or properties of visual objects that are directly linked to actions or motor performances. For instance, a handle of a cup *affords* directly a reaching and grasping action. Recently experimental studies have produced empirical evidence in support for this theory. Neuroimaging studies showed that objects activate the premotor cortex even when no action has to be performed with the object (e. g. [5,6]). Behavioural studies indicated response interferences from affordances despite the fact that they were response-irrelevant (e. g. [7,8]). For instance, a recent study by Borgh et al. demonstrated

L. Paletta and J.K. Tsotsos (Eds.): WAPCV 2008, LNAI 5395, pp. 41–53, 2009.

that pictures of hand postures (precision or power grip) can influence subsequent categorisation of objects [9]. In their study, participants had to categorise objects into either artefact or natural object. Additionally, and unknown to the participants, the objects could be manipulated with either a precision or a power grasp. Borghi et al. showed that categorisation was faster when the hand postures were congruent with the grasp compared to hand postures being incongruent with the grasp. Hence, the participants' behaviour was influenced by action-related properties of objects irrelevant to the experimental task. This experiment together with earlier, similar studies can be interpreted as evidence for an automatic detection of affordances.

Interestingly, recent experimental evidence suggests that not only actions are triggered by affordances, but also that selective attention is guided towards action-relevant locations. Using event-related potentials (ERP) Handy et al. showed that spatial attention is more often directed towards the location of tools than non-tools [10]. Pellegrino et al. presented similar evidence from two patients with visual extinction [3]. In general visual extinction is considered to be an attentional deficit in which patients, when confronted with several objects, fail to report objects on the left side of their body space. In contrast, when faced with only one object, patients can respond to the object irrespective of its location. Pellegrino et al. demonstrated that this attentional deficit can be alleviated when the handle of a cup points to the left. Pellegrino et al. interpreted their results as evidence for automatically encoded affordance (without the patients' awareness) drawing the patients' attention into their "bad" visual field.

This paper aims to lay the foundations for a computational model of such affordance-based guidance of attention. We designed a connectionist model which determines contact points for a stable grasp of an object (see Fig. 1 for an illustration). The model extracts these contact points directly from the input image. Hence, such a model could be construed as an implementation of an automatic detection of object affordances for grasping. To realise the attentional guidance through affordances, we integrated the selection mechanisms employed in the Selective Attention for Identification Model (SAIM [4]). Since this new model performs selection for action rather than identification, we termed the new model Selective Attention for Action Model (SAAM). Please note that SAAM is the first model of its kind. In this paper we will present first simulation results as well as an experimental verification of the model.

2   The Selective Attention for Action Model (SAAM)

Figure 1 gives an overview of SAAM. The input consists of black&white images. The output of the model is generated in five "finger maps" of a "hand network". The finger maps encode the finger positions which are required for producing a stable grasp of the object in the input image. At the heart of SAAM's operation is the assumption that stable grasps are generated by taking into account two types of constraints, the geometrical constraints imposed from the object shape and the anatomical constraints given by the hand. In order to ensure that the

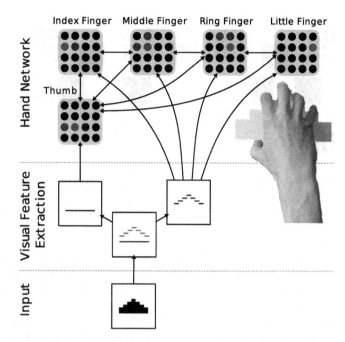

Fig. 1. Overall structure of the Selective Attention for Action Model

hand network satisfies these constraints we followed an approach suggested by Hopfield and Tank [11]. In this soft-constraint satisfaction approach, constraints define activity patterns in the finger maps that are permissible and others that are not. Then we defined an energy function for which the minimal values are generated by just these permissible activity values. To find these minima, a gradient descent procedure is applied resulting in a differential equation system. The differential equation system defines the topology of a biologically plausible network. The mathematical details of this energy minimisation approach are given in the next section. Here, we focus on a qualitative description of the two types of constraints and their implementation.

The geometrical constraints are extracted from the shape of the object in the visual feature extraction stage. To begin with, obviously, only edges constitute suitable contact points for grasps. Furthermore, edges have to be perpendicular to the direction of the forces exerted by the fingers. Hence only edges with a horizontal orientation make up good contact points, since we onlyy consider a horizontal hand orientation in this first version of the model (see Fig. 1). We implemented horizontal edge detectors using Sobel filters [12]. Finally, to exert a stable grasp, thumb and fingers need to be located at opposing sides of an object. This requirement was realized by separating the output of the Sobel filters according to the direction of the gradient change at the edge. In fact, the algebraic sign of the response differs at the bottom of a 2D-shape compared to the top of a 2D-shape. Now, if one assumes the background colour to be white and the object colour to be black, the signs of the Sobel-filter responses

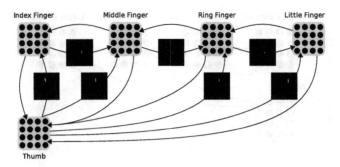

Fig. 2. Excitatory connections between fingers

indicate appropriate locations for the fingers and the thumb (see Fig. 1 for an illustration). The results of the separation feed into the corresponding finger maps providing the hand network with the geometrical constraints. Note that, of course, the assumptions about the object- and background-colours represent a strong simplification. On the other hand, this mechanism can be interpreted as mimicking the result of stereo vision. In such a resulting "depth image" real edges suitable for thumb or fingers could be easily identified.

The anatomical constraints implemented in the hand network take into account that the human hand cannot form every arbitrary finger configuration to perform grasps. For instance, the maximum grasp width is limited by the size of the hand and the arrangement of the fingers on the hand makes it impossible to place the index, middle, ring, and little finger in another order than this one. After applying the energy minimisation approach, these anatomical constraints are implemented by excitatory connections between the finger layers in the hand network (see Fig. 1 and 2). Figure 2 also illustrates the weight matrices of the connections. Each weight matrix defines how every single neuron of one finger map projects onto another finger map. The direction of the projection is given by the arrows between the finger maps. For instance, neurons in the thumb map feed their activation along a narrow stretch into the index finger map, in fact, encoding possible grip sizes. Each neuron in the target map sums up all activation fed through the weight matrices. Note that all connections between the maps are bi-directional whereby the feedback path uses the transposed weight matrices of the feedforward path. This is a direct result of the energy minimisation approach and ensures an overall consistency of the activity pattern in the hand network, since, for instance, the restriction in grip size between thumb and index finger applies in both directions. Finally, since a finger can be positioned at only one location, a winner-takes-all mechanism was implemented in all finger maps. Later in the simulation section we will show that this selection mechanism also implements global selection mimicking selective attention.

2.1 Mathematical Details

The following sections documents the mathematical details of the Selection Attention for Action Model.

Visual Feature Extraction. The filter kernel in the visual feature extraction process is a simple Sobel-filter [12]:

$$K = \begin{bmatrix} -1 & -2 & -1 \\ 0 & 0 & 0 \\ 1 & 2 & 1 \end{bmatrix} \quad (1)$$

In the response of the Sobel-filter the top edges of the object are marked with positive activation while the bottom edges are marked with negative activation. This characteristic of the filter is used to feed the correct input with the geometrical constraint applied into the finger maps and the thumb map. The finger maps receive the filter response with all negative activation set to zero. The thumb map, however, receives the negated filter response with all negative activation set to zero:

$$I_{ij}^{(\text{fingers})} = \begin{cases} R_{ij} & \text{if } R_{ij} \geq 0, \\ 0 & \text{else.} \end{cases} \quad (2)$$

$$I_{ij}^{(\text{thumb})} = \begin{cases} -R_{ij} & \text{if } -R_{ij} \geq 0, \\ 0 & \text{else.} \end{cases} \quad (3)$$

with $R_{ij} = I_{ij} * K$ whereby I_{ij} is the input image.

Hand Network. We used an energy function approach to satisfy the anatomical and geometrical constraints of grasping. Hopfield and Tank suggested this approach where minima in the energy function are introduced as a network state in which the constraints are satisfied [11]. In the following derivation of the energy function, parts of the whole function are introduced, and each part relates to a particular constraint. At the end, the sum of all parts leads to the complete energy function, satisfying all constraints.

The units $y_{ij}^{(f)}$ of the hand network make up five fields. Each of these fields encodes the position of a finger. $y_{ij}^{(1)}$ encodes the thumb, $y_{ij}^{(2)}$ encodes the index finger, and so on to $y_{ij}^{(5)}$ for the little finger. For the anatomical constraint of possible finger positions the energy function is based on the Hopfield associative memory approach [13]:

$$E(y_i) = -\sum_{\substack{ij \\ i \neq j}} T_{ij} \cdot y_i \cdot y_j. \quad (4)$$

The minimum of the function is determined by the matrix T_{ij}. For T_{ij}s greater than zero, the corresponding y_is should either stay zero or become active in order to minimize the energy function. In the associative memory approach, T_{ij} is determined by a learning rule. Here, we chose the T_{ij} so that the hand network fulfils the anatomical constraints. These constraints are satisfied when units in the finger maps that encode finger positions of anatomically feasible postures are active at the same time. Hence, the T_{ij} for these units should be greater than zero, and for all other units, T_{ij} should be less than or equal to zero. This lead to the following equation:

$$E_a(y_{ij}^{(g)}) = -\sum_{\substack{f=1 \\ g \neq f}}^{5} \sum_{g=1}^{5} \sum_{ij} \sum_{\substack{s=-L \\ s \neq 0}}^{L} \sum_{\substack{r=-L \\ r \neq 0}}^{L} T_{sr}^{(f \mapsto g)} \cdot y_{ij}^{(g)} \cdot y_{i+s,j+r}^{(f)}. \tag{5}$$

In this equation $T_{ij}^{(f \mapsto g)}$ denotes the weight matrix from finger f to finger g.

A further constraint is the fact that each finger map should encode only one position. The implementation of this constraint is based on the energy function proposed by Mjolsness and Garrett [14]:

$$E_{\mathrm{WTA}}(y_i) = a \cdot \left(\sum_i y_i - 1 \right)^2 - \sum_i y_i \cdot I_i. \tag{6}$$

This energy function defines a winner-takes-all (WTA) behaviour, where I_i is the input and y_i is the output of each unit. This energy function is minimal when all y_i are zero except one, and when the corresponding input I_i has the maximal value of all inputs. Applied to the hand network where each finger map requires a WTA-behaviour, the first part of the equation turns into:

$$E_{\mathrm{WTA}}^{a}(y_{ij}^{(f)}) = \sum_{f=1}^{5} \left(\sum_{ij} y_{ij}^{(f)} - 1 \right)^2. \tag{7}$$

The input part of the original WTA-equation was modified to take the geometrical constraints into account:

$$E_f(y_{ij}^{(f)}) = -\sum_{f=2}^{5} \sum_{ij} w_f \cdot y_{ij}^{(f)} \cdot I_{ij}^{(f)}, \tag{8}$$

$$E_t(y_{ij}^{(1)}) = -\sum_{ij} w_1 \cdot y_{ij}^{(1)} \cdot I_{ij}^{(t)}. \tag{9}$$

These terms drive the finger maps towards choosing positions at the input object which are maximally convenient for a stable grasp. The w_f factors were introduced to compensate the effects of the different number of excitatory connections in each layer.

The Complete Model. To consider all constraints, all energy functions need to be added, leading to the following complete energy function:

$$E(y_{ij}^{(f)}) = a_1 \cdot E_{\mathrm{WTA}}^{a}(y_{ij}^{(f)}) + a_2 \cdot E_{t/f}(y_{ij}^{(f)}) + a_3 \cdot E_a(y_{ij}^{(f)}). \tag{10}$$

The parameters a_i weight the different constraints against each other. These parameters need to be chosen in a way that SAAM successfully selects contact points at objects in both conditions, single-object images and multiple-object images. The second condition is particularly important to demonstrate that SAAM can mimic affordance-based guidance of attention. Moreover, and importantly, SAAM has to mimic human-style contact points. Hereby, not only the parameters a_i are relevant, but also the weight matrices of the anatomical constraints strongly influence SAAM's behaviour.

Gradient Descent. The energy function defines minima at certain values of y_i. To find these values, a gradient descent procedure can be used:

$$\tau \dot{x}_i = -\frac{\partial E(y_i)}{\partial y_i}. \tag{11}$$

The factor τ is antiproportional to the speed of descent.

In the Hopfield approach, x_i and y_i are linked together by the sigmoid function:

$$y_i = \frac{1}{1 + e^{-m \cdot (x_i - s)}}, \tag{12}$$

and the energy function includes a leaky integrator, so that the descent turns into

$$\tau \dot{x}_i = -x_i - \frac{\partial E(y_i)}{\partial y_i}. \tag{13}$$

Using these two assertions, the gradient descent is performed in a dynamic, neural-like network, where y_i can be related to the output activity of neurons, x_i the internal activity, and $\partial E(y_i)/\partial y_i$ gives the input to the neurons.

Applied to the energy function of SAAM, it leads to a dynamic unit (neuron) which forms the hand network:

$$\tau \dot{x}_{ij}^{(f)} = -x_{ij}^{(f)} - \frac{\partial E_{\text{total}}(y_{ij}^{(f)})}{\partial y_{ij}^{(f)}}. \tag{14}$$

To execute the gradient descent on a computer, a temporarily discrete version of the descent procedure was implemented. This was done by using the CVODE-library [15].

3 Study 1: Single-Object Images

The first study tested whether SAAM can generate expedient grasps in general and whether these grasps mimic human grasps. To accomplish this, simulations with single objects in the visual field were conducted. The results of the simulations were compared with experimental data on grasping these objects. In the following two sections we will at first present the experiment and its results and then compare its outcomes with the results from our simulations with SAAM.

3.1 Experiment

We conducted an experiment in which humans grasped objects. Interestingly, there are only very few published studies on this question. Most notably Carey et al. examined grasps of a stroke patient [16]. However, no studies with healthy participants can be found in the literature.

Participants. We tested 18 school students visiting the psychology department on an open day. The mean age was 17.8 years. All participants but two were right-handed. The left-handed participants were excluded from further analysis because the objects had not always been mirrored correctly during the experiment.

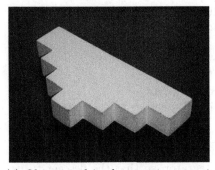

(a) Object used in the grasping experiment

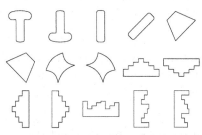

(b) Conditions of the experiment

Fig. 3. Objects and Conditions of the grasping experiment

(a) Placing of experimenter and participant

(b) Example of the pictures taken during the experiment

Fig. 4. Experimental set-up

Material. For the experiment we designed six two-dimensional object shapes. The objects were made of 2.2 cm thick wood and were painted white. Their size was between 11.5 × 4 and 17.5 × 10 centimetres (see Fig. 3a for an example). By presenting the objects in different orientations we created fifteen conditions (see Fig. 3b). Note that the shapes are highly unfamiliar, non-usable. Hence, the influence of high-level object knowledge is limited in the experiment. We chose this set-up in order to be compatible with the simulations in which SAAM possesses no high-level knowledge either.

Procedure. Figure 4a illustrates the experimental set-up. During the experiment participants and experimenter were situated on opposite sides of a glass table facing each other. The glass table was divided in two halves by a 15 cm high barrier. Participants were asked to position themselves so that their right hand was directly in front of the right half of the glass table. In each trial the experimenter placed one of the objects with both hands in the right half of the glass table. The participants were then asked to grasp the object, lift it and place it into the left half without releasing the grip. The experimenter took a picture

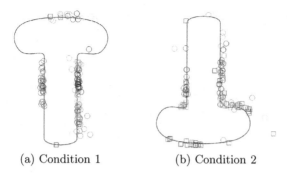

(a) Condition 1   (b) Condition 2

Fig. 5. Extracted finger positions for trials testing the first and second condition. The fingers are colour-coded: red – thumb, green – index finger, blue – middle finger, yellow – ring finger, pink – little finger.

with a camera from below the glass table (see Figure 4b for an example). After taking the photo, the participants were asked to return the object to the experimenter. The last step was introduced to ensure that the participants would not release their grasp before the photo was taken. As soon as the object was handed back to the experimenter, a new trial started by placing the next object in the right half of the glass table. Each participant took part in two blocks with fifteen trails each. The order of the trials was randomised.

Results. To analyse the pictures taken in the experiment, we developed a software for marking the positions of the fingers in relation to the objects. In Figure 5 the resulting finger positions are shown for the first and second condition. Even though the grasps show some variability, in general, participants grasped the object in two ways: they either placed their thumb at the left side of the object and the fingers on the right side or they placed the thumb at the bottom of the object and the fingers on the top edges. These two different grasping positions are indicated with two markers in Figure 5 (circle and square). Such different grasping positions were observed in all conditions.

To determine a "typical" grip from the experimental data, averaging across these very different grasping positions would not make sense. Therefore, we calculated the mean finger positions for each set of grasping positions separately. The resulting mean positions are shown in Figure 6 for all conditions. Grasping positions containing only one or two samples were discarded as outliers. For the comparison with the simulation results we only considered the grasping position for each object chosen in the majority of trials.

3.2   Simulations

We conducted simulations with SAAM using the same objects as in the experiment. Figure 7 shows two examples of the simulation results. These illustrations also include the mean finger positions from the experimental results for a comparison with the simulation data. The ellipses around the mean finger positions

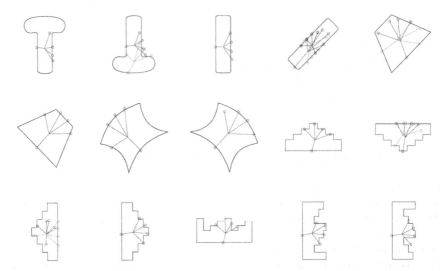

Fig. 6. Mean finger positions per class for each condition of the experiment. Different colours mark the different classes. The thumb is highlighted by a square box while the fingers are shown as circles. The individual fingers can be identified by placing the thumb of the right hand on the square box and position the fingers on the circles.

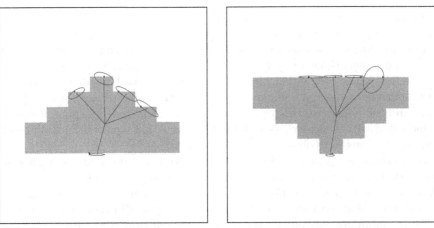

(a) Simulation 1: Experimental condition 9.

(b) Simulation 2: Experimental condition 10.

Fig. 7. Comparison of experimental results and simulated grasps. The ellipses indicate the variation in the experimental data. The black dots mark the finger positions as generated by the simulations.

illustrate the variations in the data. The comparison shows that most finger positions lie within the ellipses. Hence the theoretical assumptions behind SAAM that geometrical and anatomical constraints are sufficient to mimic human behaviour have been confirmed. Note that not all experimental conditions could be simulated with SAAM, since the model is currently only able to create horizontal grasps.

4 Study 2: Two-Object Images

SAAM produced good results for single-object displays in Study 1. This set of simulations investigated SAAM's ability to simulate attentional processes by using input images with two objects. Figure 8 shows the simulation results. The simulations are successful in the sense that contact points for only one object were selected and the second object was ignored (see Conclusion for further discussions). Note that this is an emergent property of the interplay between all constraints. The geometrical and anatomical constraints ensure that only contact points around the object were selected and the WTA-constraint restricts the contact points to one object. In addition, the weight matrices (anatomical constraints) determine the selection priorities of SAAM. At present we do not have reference data from humans. It would be especially interesting to see whether SAAM and humans have the same select preference.

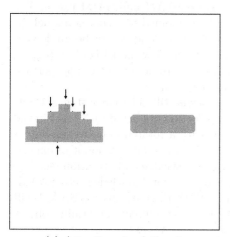 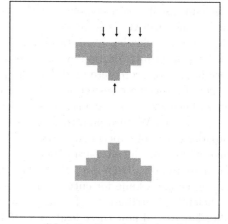

(a) Attention simulation 1.   (b) Attention simulation 2.

Fig. 8. Results for the simulation of two-object images. The black dots mark the resulting finger positions (see arrows).

5 Conclusion and Outlook

Recent experimental evidence indicates that visual attention is not only guided by visual properties of visual stimuli but also by affordances of visual objects. This paper set out to develop a model of such affordance-based guidance of selective attention. As a case in point we chose to model grasping of objects and termed the model the Selective Attention for Action Model (SAAM). To detect the parts of an object which afford a stable grasp, SAAM performs a soft-constraint satisfaction approach by means of a Hopfield-style energy minimisation. The constraints were derived from the geometrical properties of the input object and the anatomical properties of the human hand. In a comparison between simulation results and experimental data from human participants we could show that these constraints are sufficient to simulate human grasps. Note that an alternative approach would have been a complex moment analysis [17]. However, our simulations suggest that anatomical constraints render such an analysis obsolete. In a second set of simulations we tested whether SAAM cannot only extract object affordances but also implements the guidance of attention through affordances by using two-object images. Indeed, SAAM was able to select one of two objects based on their affordance. The interesting aspect here is that SAAM's performance is an emergent property from the interplay between the anatomical constraints. Especially, the competitive mechanism implemented in the finger maps is crucial for SAAM's attentional behaviour. This mechanism already proved important in the Selective Attention for Identification Model (SAIM [4]) for simulating attentional effects of human object recognition. However, it should be noted that SAAM does not select whole objects as SAIM does. Hence, SAAM's implementation of selective attention is not as intuitive as SAIM's realisation. On the other hand, since SAAM and SAIM use similar mechanisms, it is conceivable that they can be combined to form one model. In such a model SAIM's selection mechanism of whole objects can be guided by the SAAM's selection of contact points. Hence, this new model could integrate both mechanisms, selection by visual-properties and by action-related properties, forming a more complete model of selective attention.

Despite the successes reported here, this work is still in its early stages. First, we will need to verify the priorities of object selection predicted by SAAM in the second study. We also plan to include grasps with a rotated hand to simulate a broader range of experimental data. Finally, there is a large amount of experimental data on the interaction between action knowledge and attention (see [18] for a summary). Therefore, we aim to integrate action knowledge into SAAM, e. g. grasping a knife for cutting or stabbing. With these extensions SAAM will sufficiently contribute to the understanding of how humans determine object affordances and how these lead to a guidance of attention.

References

1. Gibson, J.J.: The senses considered as perceptual systems. Houghton-Mifflin, Boston (1966)
2. Gibson, J.J.: The ecological approach to visual perception. Houghton-Mifflin, Boston (1979)
3. di Pellegrino, G., Rafal, R., Tipper, S.P.: Implicitly evoked actions modulate visual selection: evidence from parietal extinction. Current Biology 15(16), 1469–1472 (2005)

4. Heinke, D., Humphreys, G.W.: Attention, spatial representation and visual neglect: Simulating emergent attention and spatial memory in the selective attention for identication model (SAIM). Psychological Review 110(1), 29–87 (2003)
5. Grafton, S.T., Fadiga, L., Arbib, M.A., Rizzolatti, G.: Premotor cortex activation during observation and naming of familiar tools. NeuroImage 6(4), 231–236 (1997)
6. Grèzes, J., Decety, J.: Does visual perception of objects afford action? evidence from a neuroimaging study. Neuropsychologia 40(2), 212–222 (2002)
7. Tucker, M., Ellis, R.: On the relations between seen objects and components of potential actions. Journal of Experimental Psychology 24(3), 830–846 (1998)
8. Phillips, J.C., Ward, R.: S–r correspondence effects of irrelevant visual affordance: Time course and specificity of response activation. Visual Cognition 9(4–5), 540–558 (2002)
9. Borghi, A.M., Bonfiglioli, C., Lugli, L., Ricciardelli, P., Rubichi, S., Nicoletti, R.: Are visual stimuli sufficient to evoke motor information? studies with hand primes. Neuroscience Letters 411(1), 17–21 (2007)
10. Handy, T.C., Grafton, S.T., Shroff, N.M., Ketay, S., Gazzaniga, M.S.: Graspable objects grab attention when the potential for action is recognized. Nature Neuroscience 6(4), 421–427 (2003)
11. Hopfield, J.J., Tank, D.W.: "neural" computation of decisions in optimization problems. Biological Cybernetics 52(3), 141–152 (1985)
12. Gonzalez, R.C., Woods, R.E.: Digital Image Processing. Addison-Wesley, Reading (1993)
13. Hopfield, J.J.: Neural networks and physical systems with emergent collective computational abilities. Proceedings of the National Academy of Sciences, vol. 79, pp. 2554–2558 (1982)
14. Mjolsness, E., Garrett, C.: Algebraic transformations of objective functions. Neural Networks 3(6), 651–669 (1990)
15. Hindmarsh, A.C., Brown, P.N., Grant, K.E., Lee, S.N., Serban, R., Shumaker, D.E., Woodward, C.S.: Sundials: Suite of nonlinear and differential/algebraic equation solvers. ACM Transactions on Mathematical Software 31(3), 363–396 (2005); also available as LLNL technical report UCRL-JP-200037
16. Carey, D.P., Harvey, M., Milner, A.D.: Visuomotor sensitivity for shape and orientation in a patient with visual form agnosia. Neuropsychologia 34(5), 329–337 (1996)
17. Mason, M.T.: Mechanics of Robotic Manipulation. Intelligent Robots and Autonomous Agents. MIT Press, Cambridge (2001)
18. Humphreys, G.W., Riddoch, M.J.: From vision to action and action to vision: a convergent route approach to vision, action, and attention. Psychology of learning and motivation 42, 225–264 (2003)

Integrating Visual Context and Object Detection within a Probabilistic Framework

Roland Perko[1], Christian Wojek[2], Bernt Schiele[2], and Aleš Leonardis[1]

[1] University of Ljubljana, Slovenia
{roland.perko,ales.leonardis}@fri.uni-lj.si
[2] TU Darmstadt, Germany
{wojek,schiele}@cs.tu-darmstadt.de

Abstract. Visual context provides cues about an object's presence, position and size within an observed scene, which are used to increase the performance of object detection techniques. However, state-of-the-art methods for context aware object detection could decrease the initial performance. We discuss the reasons for failure and propose a concept that overcomes these limitations, by introducing a novel technique for integrating visual context and object detection. Therefore, we apply the prior probability function of an object detector, that maps the detector's output to probabilities. Together, with an appropriate contextual weighting, a probabilistic framework is established. In addition, we present an extension to state-of-the-art methods to learn scale-dependent visual context information and show how this increases the initial performance. The standard methods and our proposed extensions are compared on a novel, demanding image data set. Results show that visual context facilitates object detection methods.

1 Introduction

A standard approach for detecting an object of a known category in still images is to exhaustively analyze the content of image patches at all image positions and at multiple scales (see e.g. [1,2]). When a patch is extracted from an image, it is classified according to its local appearance and associated with a detection score. The score should correspond to the probability of the patch representing an instance of the particular object category and is usually mapped to a probability score. As it is known from the literature on visual cognition [3,4], cognitive neuroscience [5,6] and computer vision [7,8,9], the human and animal visual systems use relationships between the surrounding and the objects to improve their ability of categorization. In particular, visual context provides cues about an object's presence, position and scale within the observed scene or image. This additional information is typically ignored in the object detection task. Like in other promising papers on visual context for object detection [10,8,11,9], we define the context as the surrounding, or background, of the current object of interest. This context is used to focus the attention on regions in the image where the objects are likely to occur. Instead of searching the whole image at various

L. Paletta and J.K. Tsotsos (Eds.): WAPCV 2008, LNAI 5395, pp. 54–68, 2009.

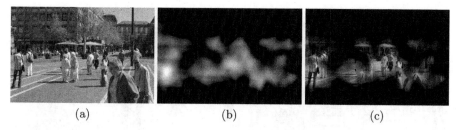

(a)   (b)   (c)

Fig. 1. A concept of using visual context for object detection. (a) A standard image of an urban scene, (b) the focus of attention for the task of pedestrian detection using the method in [9] and (c) the image multiplied by the focus of attention.

scales for an object, visual context provides regions of interest, i.e. the focus of attention, where the search is restricted to. This results in speedup, but more importantly, it can increase the detection rate by not considering incorrect object hypotheses at unlikely positions in the first place. This concept is illustrated in Fig. 1. In this study, we conducted experiments with the approaches of Hoiem *et al.* [8] and Perko and Leonardis [9] on a specific task of detecting pedestrians in urban scenes. We found that the first approach may fail when too many incorrect object hypotheses are detected and the second may, in some cases, reduce the detection rate of the original detector. A detailed analysis revealed that these failure cases are linked to the assumption of these methods that visual contextual information *always* assists the detection step. Furthermore, the prior probability from the local appearance-based object detection method is ignored when combined with the contextual score. We state that the prior probability is an intrinsic property of the object detector used and is defined as the conditional probability of the detection being correct given the detection score. The function is used to map the detector's output to a probability space. In addition, the contextual information is weighted by a function depending on the probabilistic detection score. The basic idea is as follows: if an object is well defined by its local appearance, then context should not contribute much in the detection phase. It can even introduce additional errors by incorrectly re-ranking the detections. However, if the local appearance is weak, context can contribute significantly to improve detections. Therefore, we propose to learn this prior probability function, together with a contextual weighting, and embed it into the existing systems for object detection. An example of this concept is given in Fig. 2. The pedestrian in Fig. 2(b) is well defined by its local appearance and context is not important to get an unambiguous detection. However, the smaller pedestrians shown in Fig. 2(c) and (d) are not as easily detected based on local appearance alone, so that visual context provides more clues about these detections being correct.

Our contribution. We point out in which situations the current state-of-the-art methods for performing context aware object detection decrease the initial performance in practice and discuss the reasons for failure. In particular the approaches in [8] and [9] are evaluated and compared. We then propose a concept

(a)   (b)   (c)   (d)

Fig. 2. A standard image of an urban scene. (a) Three pedestrians of different size are marked by the yellow arrows and boxes. (b-d) Close-ups of the marked objects using nearest-neighbor interpolation for resizing. Even though humans can recognize the objects at all three scales, it is obvious that the bigger pedestrian in (b) is easier to detect than the ones in (c) or (d).

that overcomes these limitations. More specifically, the contextual information is combined with the local appearance-based object detection score in a fully probabilistic framework. We also extend this framework to multiple object detectors trained for different object sizes. In addition, we present an extension to the method in [9] to learn scale-dependent visual context information and show how its performance increases. Then, the methods are compared on a novel demanding database.

Organization of the paper. Related work will be discussed in detail in Sec. 2 and the drawbacks of the state-of-the-art approaches are pointed out in Sec. 3. In Sec. 4 we describe the extension to standard methods and how the probabilistic framework is set up. Results are given in Sec. 5. In the discussion in Sec. 6 we analyze the limitations of contextual processing and conclude the paper in Sec. 7.

2 Related Work

In computer vision the combination of visual context with the task of object detection is a rather young field of research. The aim is to extract more global information from a single image and use it to improve the performance of classical object detection methods. Technically there are two issues to be solved. First, how to represent this kind of visual context within some data structure, i.e. a feature vector. There is a lack of simple representations of context and efficient algorithms for the extraction of such information from images [12]. And second, how to combine this information with an object detection technique. For the former the feature vectors holding contextual information are learned from a labeled database. The LabelMe image database [13] is often used for such purposes. Then for new images, the extracted feature vectors are classified using this learned model. For the latter it is assumed that the contextual information and the local information used for detecting objects are statistically independent.

Therefore, their conditional probability is equal to the product of the individual probabilities [10,9]. Namely, the combined probability p is calculated using the contextual probability p_C and the local appearance-based probability p_L as $p = p_C \cdot p_L$. Context could be also used in a cascade [11,9]. In this case, only pixels with a contextual confidence above a threshold are used in the object detection task. However, as the detection score is not re-ranked and therefore only out-of-context detections (e.g. pedestrians in the sky) are filtered, the increase of detection rate is negligible [14]. Before we present three methods in detail we want to point to the current review on visual context and its role in object recognition by Oliva and Torralba [12]. They also show how the focus of attention extracted using visual context can be combined with classical attention concepts, e.g. with the system of Itti and Koch [15]. Therefore, the results from visual context can be used as the top-down saliency in approaches like [16].

The influential work from Oliva and Torralba, e.g. [17,18,19,10], introduced a novel global image representation. The image is decomposed by a bank of multi-scale oriented filters, in particular four scales and eight orientation. The magnitude of each filter is averaged over 16 non-overlapping blocks in a 4×4 grid. The resulting image representation is a 512-dimensional feature vector, which is represented by the first 80 principal components. Despite the low dimensionality of this representation, it preserves most relevant information and is used for scene categorization, such as a landscape or an urban environment. Machine learning provides the relationship between the global scene representation and the typical locations of the objects belonging to that category. To the best of our knowledge there exist no evaluation for the combination of this derived focus of attention with a state-of-the-art object detection algorithm. In a real scenario a coarse prior for the possible object location in the image does not automatically increase the performance of an object detector. As seen later, when combined just by multiplication the results of the detection may and often do degrade.

Hoiem et al. provided a method to extract the spatial context of a single image [20]. The image is first segmented into so called superpixels, i.e. a set of pixels that have similar properties. These regions are then described by low level image features, i.e. color, texture, shape and geometry, forming a feature vector. Each region is classified into a semantic class, namely ground, vertical and sky, using a classifier based on AdaBoost with weak decision tree classifiers. As a result each pixel in the input image is associated with the probabilities of belonging to these three classes. For the task of object detection this classification provides useful cues and they are exploited in [8] and [9]. Hoiem et al. [8] use the coarse scene geometry to calculate a viewpoint prior and therefore the location of the horizon in the image. The horizon, being the line where the ground plane and the sky intersect in infinity, provides information about the location and sizes of objects on the ground plane, e.g. pedestrians or cars. The scene geometry itself limits the location of objects on the ground plane, e.g. no cars behind the facade of a building. Now, the innovative part of their work is the combination of the contextual information with the object hypotheses using inference. Without going into detail, the main idea is to find the object hypotheses that are consistent in terms of size and location, given the geometry and horizon of the scene. As a result, a cluster of object hypotheses is determined, that fits the data best. This contextual inference uses the global visual context and the relation between objects in that scene. The position of the horizon is an integral part of this system, limiting the approach to object categories that are placed on the ground plane and to objects of approximately the same size. E.g. the approach cannot be used to detect windows on facades or trees.

Perko and Leonardis [9] use the semantic classification of an image in [20] as one feature set, and low-level texture features, based on Carson's *et al. Blobworld* system [21] as a second set. Both types of features are probabilistic and extracted for each pixel in the image, which is downsampled for speedup. To define the visual context at a given position in the image, they sample those features radially for a given number of radii and orientations, like in [11]. The extracted feature vector is relatively low-dimensional, i.e. 180-dimensional as reported in [9]. A significant increase of the object detection rate is reported using this kind of contextual information, where the low-level texture-based scene representation is more important than the high-level geometry-based representation.

3   Drawbacks of the State-of-the-Art Approaches

The mentioned state-of-the-art methods for extracting and using visual context for an object detection task are reported to increase the performance of the initial object detection. However, we found that this is not always the case, especially when using a demanding image database. By *demanding* we mean images with a lot background clutter and textured regions where object hypotheses are often incorrect, and where objects occur at very different scales. We therefore collected an image data set in an urban environment and experimented with the methods in [8] and [9], using pedestrians as objects of interest. As done in the mentioned papers we plotted the detection rate versus the false positives per image (FPPI), and observed that the method by Hoiem *et al.* significantly decreased the initial detection rate. In the evaluation we used the publicly available Matlab source code[1]. Fig. 3(a) shows the initial detection rate using our own implementation of the Dalal and Triggs pedestrian detector [2] and the detection rate curves after applying the inference. Our analysis shows that there are two reasons for this behavior. First, the contextual inference process often produces an incorrect solution. A cluster of object hypotheses is determined that satisfies a viewpoint estimate which is however incorrect. In such a case typically *all* detections in that image are incorrect and correct detections are mostly discarded. An example is given in Fig. 3(b-c). In our database this happens for 10.1% of the images. We consider the horizon estimate as correct if its position w.r.t. the ground truth horizon position deviates maximal 10% of the image's height. Second, in this approach the object detection score is assumed to be probabilistic. Therefore, the support vector machine (SVM) outputs from object detection are mapped to probabilities using the approach in [22]. However, these mapped outputs are

---

[1] http://www.cs.uiuc.edu/homes/dhoiem/software/

only probabilistic in the sense, that they are in the range of $[0,1]$. The relation from these scores to the probability that a detection is correct is still unknown. This additional mapping (called the prior probability) should be learned from training data and used to have real probabilities in this framework. As shown in Sec. 5.2 the method performs better when incorporating this function. The method of Perko and Leonardis [9] does not have problems when many incorrect hypotheses are given, as the detections are treated separately (no contextual inference). However, as the prior probability function is not modeled, locally well defined objects could be incorrectly re-ranked. In addition, the contextual influence is not specially related to the appearance-based score. Due to this two aspects the method performs poor at low FPPI rates, yielding even worse results than the initial detections not using contextual information at all. Fig. 3(a) gives the detection rate curves. In general we noticed that the current state-of-the-art methods for performing context aware object detection could fail in practice. We can also predict that the methods of Torralba [10] and Bileschi [11] will likewise lower the initial detection rate, as they are ignoring the prior object detection probability as well, and the underlying concept is similar to [9]. Using the prior probability from the object detector will fix these problems for all mentioned methods.

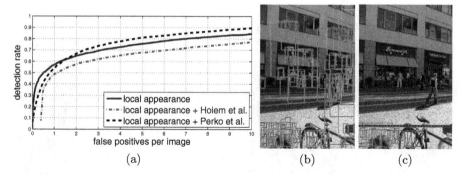

(a)   (b)   (c)

Fig. 3. Drawbacks of the state-of-the-art approaches that use visual context for object detection. (a) The initial detection rate based on local appearance only yield better results than the two evaluated methods that use contextual information. The approach in [8] decreases the performance in general, while the one in [9] has problems at low FPPI only. (b) An example image with object hypotheses from pedestrian detection. (c) Due to the cluster of incorrect hypotheses in the foreground the horizon estimate (horizontal line) is invalid, so that all detections after contextual inference are incorrect.

4   Our Extensions

Two extension to the existing methods are presented. First, we introduce a scale extension in Sec. 4.1 and second, we explain the probabilistic framework in Sec. 4.2.

4.1   Scale Extension

We implemented an extension to [11] and [9], i.e. to learn and use a scale-dependent visual context information. The mentioned methods use a fixed sized region, i.e. one quarter of the image area, to extract the feature vector holding the contextual information. However, when dealing with objects of an a priori known range of sizes in the real world, e.g. pedestrians or cars, these fixed sized regions are not representing the same semantic context for objects perceived at different scales. Therefore, these regions should be scaled with the object's size in the image. Smaller objects corresponds to objects in the distance, an effect of the projective geometry. Therefore, we simply scale the region of which the context is gathered for the given object of interest with its size, visualized in Fig. 4. For smaller pedestrians a smaller region is used to extract the features. These new features are learned using an SVM as in [9]. Then, instead of extracting only one context confidence map for one predefined size, multiple confidence maps are calculated for a given number of scales. An example of such confidence maps is given in Fig. 4 for six scales. The context confidence score for a given detection is then calculated by linear interpolation using the scores of the two adjacent scales. As a result the prior based on context is not only able to provide regions where an object is likely to occur, it also provides the possible size of the object. It

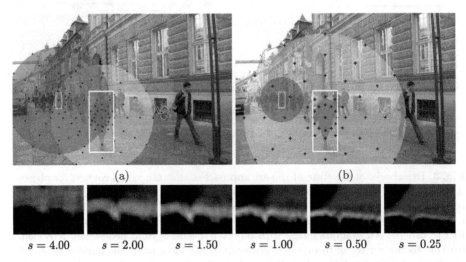

$s = 4.00$ $s = 2.00$ $s = 1.50$ $s = 1.00$ $s = 0.50$ $s = 0.25$

Fig. 4. Top row: The regions from which the contextual information is gathered in a feature vector is visualized with the red and yellow circles for the two marked objects. The blue crosses indicate the locations where the contextual information is sparsely sampled. (a) The regions are of constant size as proposed in [11,9]. (b) Regions are scaled according to the object's size so that they represent similar semantic information. Bottom row: Context confidence maps (foci of attention) based on geometry features for six scales s for the image in the top row. Bright regions indicate locations where pedestrians are likely to occur. It is visible that smaller objects are more likely to occur at different locations than bigger objects.

should be pointed out that extracting several foci of attention of course increase the computational expanses of the whole framework. However the confidence map extraction using the pre-learned SVM is faster than the feature extraction, so that the multiscale approach is also applicable in practise.

4.2 Probabilistic Framework

As stated before the prior probability of the object detector used should be modeled and applied to the whole framework. This probability function is an intrinsic property of the detector and can be learned in the training phase. It holds the conditional probability of the detection being correct given the detection score, which is in the probabilistic range. This function is only valid for one set of parameters, that means if, e.g., a detection threshold is changed the function has to be recalculated. In our approach we label the detections as true positives and false positives using the ground truth that exists in the learning phase of object detection. Two histograms with 16 bins are calculated holding the number of true and false detections. The prior probability is then extracted by dividing the number of true detections by the number of overall detections in each bin. To ensure a smooth function the values are filtered using average and median filtering, where the borders, i.e. values at 0 and 1 are preserved. Then an analytic function p_a is fitted describing the prior probability. In the current implementation a polynomial of order 6 is used. The concept is illustrated in Fig. 5(a-b). Instead of multiplying the local appearance-based detection score L with the contextual score p_C at the given position in the image as in [10,11,9], the final score is the product of the prior probability of the detection being correct $p_a(L)$ with the context confidence p_C weighted by the function w, defined as

$$p_{combined} = p_a(L) \cdot w \cdot p_C \quad \text{with} \quad w = (1 - p_a(L))^k + 1. \qquad (1)$$

The parameter k defines the steepness of the weighting function w, where we use $k = 2$ in the experiments. We experimented with different values of k and

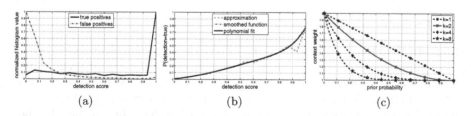

(a)   (b)   (c)

Fig. 5. Extraction of the prior probability function illustrated using Dalal and Triggs pedestrian detector [2] in (a-b) and the function used to weight the influence of the contextual information in (c). (a) Normalized histograms for true positives and false positives, (b) approximation of the conditional probability function, smoothed prior probability function and the polynomial fit p_a. (c) Function used to weight the influence of the contextual information, shown for different setting of the parameter k. Context gets a higher weight for detections with lower prior probabilities.

found out that the results improve when using the contextual weighting, i.e. $k \neq 0$. However, the specific value of k in rather unimportant as long as it is the range of $[1, 4]$. The function w is bounded in $[1, 2]$ and visualized in Fig. 5(c). This weighting models the concept that the contextual information gets higher weight for detections with a lower local appearance-based score, and lower weight for high-ranked detections.

5   Experimental Results

We conducted experiments using the proposed extensions described in Sec. 4 and show that the methods in [8] and [9] yield significantly better results with these extensions. First, we evaluate the scale extension in Sec. 5.1 and second, we analyze the probabilistic framework in Sec. 5.2. Additional results including videos can be found on our project page[2].

5.1   Scale Extension

To evaluate the results of the scale extension to [9] (see Sec. 4.1), we used the same data set as in the original paper, the Ljubljana urban image data set[3], and compared them to the initial results. The first result is, that the positive feature vectors are more similar (smaller standard deviation) compared to the original method. This indicates that the contextual information grasps a more similar semantic representation, when scaling the regions according to the object's size. The second result is an increase of the detection rate. Using the Seemann *et al.* detector [23] the increase is 4.1%, i.e. a relative increase of 19.2% over the original method. For the detections using Dalal and Triggs detector [2] the increase is 1.3%, i.e. relative increase of 22.3%. These numbers are calculated at a fixed rate of 2 FPPI. Fig. 6(a) shows the detection curves for the original approach and using the scale extension for the Seemann *et al.* detections. Fig. 6(b) visualizes the contributions of the three contextual cues to the final result, where the cue based on texture benefits most using the scale extension.

5.2   Probabilistic Framework

Darmstadt urban image data set. To test the new framework we collected a demanding image data set containing 1572 images of the city of Darmstadt[4] with a resolution of 1944×2896 pixels each. The images were downsampled to 972×1448 pixels for our evaluation. 4133 pedestrian were manually labeled and used as ground truth in the experiments. Each pedestrian is defined by the corresponding bounding box, where the whole object is inside. The bounding boxes have a fixed aspect ratio of $1 : 2$, centered at the object. For the small scale object detection task a subset of 121 images were taken (each 13th image) and

---

[2] http://vicos.fri.uni-lj.si/research/visual-context/

[3] http://vicos.fri.uni-lj.si/luis34/

[4] http://vicos.fri.uni-lj.si/duis131/

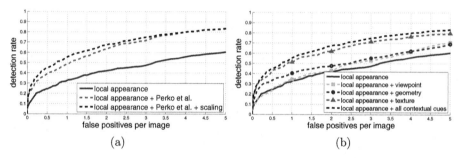

Fig. 6. Detection rate curves for the scale extension. (a) Comparison of the original approach of [9] with our proposed scale extension. At 2 FPPI the original method boosted the performance by 21.3%, while the new approach increased it by 25.4%, an relative increase of 19.2%. At lower FPPI rates the boost is even more significant. (b) Contribution of the three contextual cues to the final result.

all pedestrians were labeled, down to 12×24 pixels, resulting in 661 pedestrians. This subset is called *sub13* in the rest of the paper.

Object detectors. As seen in Sec. 5.1 and also addressed in [9] a weak object detector can easier be boosted using contextual information than a detector which gives very good results in the first place. As we aim for the more difficult task, we show that even the results of the best detectors can be significantly improved using visual context. Therefore, we use a detector based on Dalal and Triggs [2], which is one of the best pedestrian detectors currently available.

Prior probabilities of the object detectors. Fig. 7(a) shows the prior probabilities for the detectors in [2] and [23]. It is obvious, that the initial detection scores are rather different from the observed probabilities and that the prior probabilities vary for the two detectors. To experimentally prove our claim, that smaller objects are more difficult to detect than larger objects, we trained our own version of a *histogram of gradients (HoG)* based detector for different object sizes. Four detectors are trained for 24×48, 32×64, 40×80 and 64×128 pixels. Each detector then collects detections within its range, i.e. the first one collects detections with a height from 1 to 63 pixels, the next from 64 to 79 pixels and so forth. The four prior probability functions are shown in Fig. 7(b). As expected, the probability of a detection being correct is higher for larger objects. For objects larger than 128 pixels in height a detection score of 1 indicates that the detection is correct with 89%, while for smaller objects up to 63 pixels the same score indicates a correctness of only 59%. These prior probabilities are used to re-rank the detections. Like above, the initial detection score is quite different from the observed probabilities. For example scores up to 0.5 only indicate a true detection with less than 10% (see Fig. 7(b)). Therefore, all algorithms which take the initial detector's score within a probabilistic framework yield inaccurate results. The new detection score which corresponds to the observed probabilities is $p_a(L)$ and should be used in the approaches in [10,8,11,9].

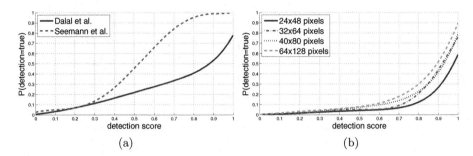

Fig. 7. Prior probability functions given for different detectors. (a) For Dalal and Triggs detector [2] and for Seemann *et al.* detector [23] and (b) for our HoG-based detector trained for four different object sizes.

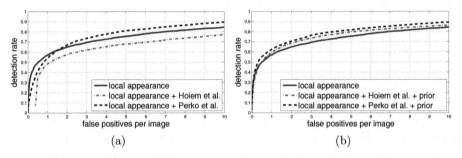

Fig. 8. Detection rate curves. (a) Plotted for the original approach of [8] and [9] and (b) for our proposed extensions using the prior probabilities and contextual weighting. While the original methods decrease the accuracy, they yield good results when incorporating the proposed extensions.

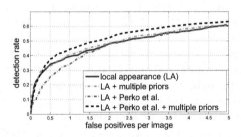

Fig. 9. Detection rate curves plotted for the *sub13* data set. Using multiple prior functions increase the performance of the local appearance-based object detection and of contextual inference. The detection rate is scaled to 0.7.

Results using visual context. The original results together with the results using our proposed extensions are given in Fig. 8 for [8] and [9]. With our extensions included both methods are able to increase the initial object detection performance, with an average boost of about 4% for [8] and 7% for [9]. Similar results are achieved using the *sub13* data set using multiple prior functions, one for each trained detector. Results are given in Fig. 9, where we compare the initial detector's performance with our concept of multiple prior functions, once with and without using contextual inference. As expected the original method [9] performs poorly, while the performance increases when incorporating these probabilities.

6   Discussion

With our extension presented in Sec. 4 the two methods [8,9] for performing visual context aware object detection are improved. However, the increase of the detection rate is on average only about 4% for [8] and 7% for [9] (see Fig. 8). Depending on the final application this boost is of interest or may be negligible. An interesting question is why these novel methods are not providing stronger cues to assist object detection. Part of the answer is illustrated within Fig. 10. In (a) two images of our data set are shown with all object hypotheses marked, and in (b) all detections with a score $p_a(L) > 0.5$. In (c) the horizon estimate from [8] is visualized with the remaining object hypotheses after contextual inference. Even though the horizon is correctly estimated and all 11 (top row) respectively 9 (bottom row) detections satisfy the global scene geometry, only 1 of them is a correct detection in each row. In (d) the location priors from [9] are shown for geometry features (shown for the scale $s = 1$, cf. Fig. 4). These priors are robust estimates, however they will only down-rank a few detections with a high score, i.e. the hypothesis on the roof top in the second example. In general the problem is that there are many object hypotheses based on a local appearance measure that are incorrect and suit to the scene in terms of their position and size. Such hypotheses cannot be rejected or down-ranked by visual contextual information. Another aspect is the way how the contextual information is integrated with local appearance-based object detection. In Eq. (1) the prior probability of the object detector and a contextual weighting is introduced. However, the dependencies of the individual contextual scores and the object detection score are not modeled. Therefore, the next step would be to estimate the conditional probability density function of all cues, which could then be used to increase the overall performance.

To put it in simple words: Visual context only provides priors for the position and size where an object of interest is likely to occur according to the given scene content. On the one hand, false object hypotheses fitting to the scene layout "survive" the contextual inference. On the other hand, hypotheses that are strongly out-of-context have weak local appearance in many cases. Due to this aspects, the boost of the detection rate is limited using visual context as an additional cue. However, we assume that computer vision researcher will come up with more robust object detection methods based on local appearance. Visual context could then be used to prune the few out-of-context hypotheses with high detection score and to limit the search space for the detection.

(a)   (b)   (c)   (d)

Fig. 10. Limits of visual context aware object detection. (a) Urban scene with hypotheses for pedestrians, (b) object hypotheses with a score larger than 0.5, (c) horizon estimate and detections supporting this estimate [8] and (d) focus of attention using geometry features [9]. Best viewed in color.

7 Conclusion

Visual context provides cues about an object's presence, position and size within the observed scene, which are used to increase the performance of object detection techniques. However, state-of-the-art methods [8,9] for context aware object detection could decrease the initial performance in practice, where we discussed the reasons for failure. We proposed a concept that overcomes the limitations, using the prior probability of the object detector and an appropriate contextual weighting. In addition, we presented an extension to state-of-the-art methods [11,9] to learn scale-dependent visual context information and showed how this increases the initial performance. The methods and our proposed extensions were compared on a novel demanding database, where the object detection rate was increased by 4% to 7% depending on the method used.

Acknowledgements. We thank all the authors who made their source code or binaries publicly available, so that we avoided painful re-implementation. In particular we thank Derek Hoiem, Navneet Dalal and Edgar Seeman. This research has been supported in part by the following funds: Research program Computer Vision P2-0214 (RS), EU FP6-511051-2 project MOBVIS and EU project CoSy (IST-2002-004250).

References

1. Viola, P., Jones, M.: Rapid object detection using a boosted cascade of simple features. In: Proc. Conf. Comp. Vis. Pattern Recog. (December 2001)
2. Dalal, N., Triggs, B.: Histograms of oriented gradients for human detection. In: Proc. Conf. Comp. Vis. Pattern Recog., vol. 2, pp. 886–893 (June 2005)
3. Palmer, S.E.: The effects of contextual scenes on the identification of objects. Mem. Cognit. 3, 519–526 (1975)
4. Biederman, I.: Perceptual Organization. In: On the semantics of a glance at a scene, pp. 213–263. Lawrence Erlbaum, Mahwah (1981)
5. Bar, M.: Visual objects in context. Nat. Rev. Neurosci. 5, 617–629 (2004)
6. Aminoff, E., Gronau, N., Bar, M.: The parahippocampal cortex mediates spatial and nonspatial associations. Cereb. Cortex 17(7), 1493–1503 (2007)
7. Torralba, A., Oliva, A., Castelhano, M.S., Henderson, J.M.: Contextual guidance of attention in natural scenes: The role of global features on object search. Psychol. Rev. 113(4), 766–786 (2006)
8. Hoiem, D., Efros, A.A., Hebert, M.: Putting objects in perspective. In: Proc. Conf. Comp. Vis. Pattern Recog., vol. 2, pp. 2137–2144 (June 2006)
9. Perko, R., Leonardis, A.: Context driven focus of attention for object detection. In: Paletta, L., Rome, E. (eds.) WAPCV 2007. LNCS, vol. 4840, pp. 216–233. Springer, Heidelberg (2007)
10. Torralba, A.: Contextual priming for object detection. Int. J. Comput. Vision 53(2), 153–167 (2003)
11. Bileschi, S.M.: StreetScenes: Towards Scene Understanding in Still Images. PhD thesis, Massachusetts Institute of Technology (May 2006)
12. Oliva, A., Torralba, A.: The role of context in object recognition. Trends in Cognit. Sci. 11(12), 520–527 (2007)
13. Russell, B.C., Torralba, A., Murphy, K.P., Freeman, W.T.: LabelMe: A database and web-based tool for image annotation. Technical Report AIM-2005-025, MIT AI Lab Memo (September 2005)
14. Wolf, L., Bileschi, S.M.: A critical view of context. Int. J. Comput. Vision 69(2), 251–261 (2006)
15. Itti, L., Koch, C.: Computational modeling of visual attention. Nat. Rev. Neurosci. 2(3), 194–203 (2001)
16. Rasolzadeh, B., Targhi, A.T., Eklundh, J.O.: An Attentional System Combining Top-Down and Bottom-Up Influences. In: Paletta, L., Rome, E. (eds.) WAPCV 2007. LNCS, vol. 4840, pp. 123–140. Springer, Heidelberg (2007)
17. Oliva, A., Torralba, A.: Modeling the shape of the scene: A holistic representation of the spatial envelope. Int. J. Comput. Vision 42(3), 145–175 (2001)
18. Torralba, A., Sinha, P.: Statistical context priming for object detection. In: Proc. Int. Conf. Computer Vision, vol. 1, pp. 763–770 (July 2001)
19. Torralba, A.: Contextual modulation of target saliency. In: Neural Inf. Proc. Systems, vol. 14, pp. 1303–1310 (2002)

20. Hoiem, D., Efros, A.A., Hebert, M.: Geometric context from a single image. In: Proc. Int. Conf. Computer Vision, vol. 1, pp. 654–661 (October 2005)
21. Carson, C., Belongie, S., Greenspan, H., Malik, J.: Blobworld: Image segmentation using expectation-maximization and its application to image querying. IEEE Trans. Patter. Anal. Mach. Intell. 24(8), 1026–1038 (2002)
22. Platt, J.C.: Probabilistic outputs for support vector machines and comparisons to regularized likelihood methods. Advances in Large Margin Classifiers 10(3), 61–74 (1999)
23. Seemann, E., Leibe, B., Schiele, B.: Multi-aspect detection of articulated objects. In: Proc. Conf. Comp. Vis. Pattern Recog., vol. 2, pp. 1582–1588 (June 2006)

The Time Course of Attentional Guidance in Contextual Cueing

Andrea Schankin and Anna Schubö

Ludwig Maximilian University Munich, Department Psychology,
Leopoldstr. 13, 80802 Munich, Germany
schankin@psy.lmu.de, anna.schuboe@lmu.de

Abstract. Contextual cueing experiments show that targets in heteroge-
neous displays are detected faster with time when displays are repeated,
even when observers are not aware of the repetition. Most researchers
agree that the learned context guides attention to the target location and
thus speeds subsequent target processing. Because in previous experiments
one target location was uniquely associated with exactly one configuration,
the context was highly predictive. In two experiments, the predictive value
of the context was investigated by varying the number of possible target
locations. We could show that even when the context was less predictive, it
was learned and used to guide visual-spatial attention. However, the time
course of learning differed significantly: learning was faster when the number of target locations was reduced. These results suggest that not an association of context and target is learned but that rather the precision of
the attention shift improves.

Keywords: Contextual cueing, learning, perception, visual attention,
cognitive systems.

1 Introduction

In real-world, objects usually occur in a relatively constant spatial context with
stable relationships between the context elements. For example, we are used to
find some specific objects on our desk, e.g., a computer monitor, a keyboard,
a mouse, staples of paper and a coffee cup. By knowing the locations of any
set of objects, e.g. the objects on the table in the previous example, one often
also knows (or can at least predict) the location of a single target, e.g. the
coffee cup, thereby reducing or eliminating the need to execute a detailed serial
search throughout the entire scene. A stable, meaningful scene structure may
thus be used to help guiding visual attention to behaviorally relevant targets
and may serve to constrain visual processing. In a series of studies Chun and
Jiang [1] demonstrated that if the target item was embedded in an invariant
configuration that was repeated across the experiment, reaction times (RTs) to
find the target were faster than when the target item appeared in a novel or
unrepeated configuration. This effect has become known as *contextual cueing*.

To investigate contextual cueing effects in the laboratory, participants are
usually asked to search through a display of objects to identify a target object.

L. Paletta and J.K. Tsotsos (Eds.): WAPCV 2008, LNAI 5395, pp. 69–84, 2009.

In the example described above, people would be presented with the picture of a table and would be asked to look for the coffee cup. The variable of interest, the context, is defined by the spatial arrangement of the irrelevant (distracting) objects. For example, if the coffee cup is the object to be found (target), the monitor, a staple of paper, pens etc. are distracting objects. The arrangement of these distracting objects on the table, i.e. how the monitor is related to the keyboard, to the staple of paper, and to the pens, forms the visual context. In a typical contextual cueing experiment, a set of these displays are repeated throughout the entire experimental session, meaning a picture of this table is presented repeatedly. These displays are referred to as OLD context. In a control condition, NEW context displays, which are randomly generated in each block, e.g. other pictures of differently-looking tables and objects on it, are presented intermingled with OLD context displays. It is important to note here that, on a particular repeated picture, the target object is always placed at the same location.

In this kind of experiments, usually two learning processes occur: general and context-specific learning. *General learning* is reflected by decreasing reaction times in the time course of the experiment. That means, finding the coffee cup becomes easier (and faster) independent of the context it is presented in. This effect is probably due to training effects, e.g., habituation to the sensory input, allocation of attentional resources and speeded response selection mechanisms. By general learning, RTs to targets presented in OLD and NEW contexts are decreased as well. In addition to this general learning, *context-specific learning* occurs after four to five repetitions of the OLD displays; RTs are faster in the OLD as compared to those in the NEW context condition. This effect, that has become known as *contextual cueing*, shows that participants were probably encoding the context information, even though they were not told it was informative. In reference to the example described above, one is faster in finding the coffee cup on a familiar table than on a new one.

Most researchers agree that, in the time course of the experiment, an association between the context and the target location is formed and afterwards used to guide attention to the target location [2, 3]. Importantly, the acquisition of contextual knowledge is supposed to be implicit. That is, contextual knowledge is acquired passivly and incidentally. This kind of learning allows complex information about the stimulus environment to be acquired without intention or awareness [1]. That is, we are faster in finding the coffee cup on a familiar table because we have learned where it is probably placed.

As described above, the association between the visual context and the target location is a necessary precondition of contextual cueing. In a typical contextual cueing experiment, this association is formed after four to five repetition of a particualar search display. Previous experiments have investigated different factors which may influence the learning process. Jiang and Chun [4], for example, have shown that the context has to be attended in order to obtain a contextual cueing effect. In their experiments, participants performed a visual search through context items presented in an attended color and in a to-be-ignored color. Contextual cueing only occurred when the repeated context was attended. It has also been shown that the contextual cueing effect depends on the task affordances. In an experiment by Endo and Takeda [5], participants learned the spatial configurations of the contexts but not the object identities, even when both configurations and identities were completely correlated. On the other hand, when only object identities (or arrangements) were repeated, an effect of identity (or arrangement) learning could be observed. Furthermore, additive effects of configuration learning and identity learning were observed when, in some trials, each context was the relevant cue for predicting the target. Thus, contextual learning occurs selectively, depending on the predictability of the target location.

To shortly summarize these results, the learned association of context and target location is affected by the predictive value of the context. As discussed previously, the context is defined as the arrangement of distracting objects. In the experiments cited above, only those objects (as part of the context) were associated with the target location which were either attended [4] or task-relevant [5]. Although the information content of the whole context was reduced (i.e., only some of the objects or some features of the objects delivered information to predict the target location), these objects, once dectected, predicted the target location reliably. That is, although the information content of the context was reduced in general, its value to predict the target location and to guide attention was still high in order to solve the required task efficiently.

In this paper we want to investigate the role of information content on contextual cueing further. Does contextual cueing occur when the context is less predictive for the target location?

2 Search Strategies in Contextual Cueing

In typical contextual cueing experiments, participants are presented with search displays and asked to find a target among distractors. Assuming that there are 24 possible target locations in each trial, attention has to be shifted serially to each of them until the relevant object is found. Thus, the probability to find the target just by chance is quite low, namely 1:24. In these experiments, usually one specific context (configuration) is associated with exactly one target location. For example, if there were 24 possible target locations, 12 randomly chosen locations were associated with 12 OLD displays, and the remaining 12 locations were used in NEW configurations [1, 2]. If the visual system noticed that the context is OLD vs. NEW, the probability to find the target would be already reduced to 1:12. Thus, the context is highly informative to predict the location of the target. However, does contextual cueing still occur when the context is less predictive?

To investigate the effect of information content, we varied the number of target locations while the number of configurations was kept constant. Thereby, we were able to compare the size of the contextual cueing effect of a 'classical' setting, in which each target location is associated with exactly one configuration and in which the guidance of attention is high (Exp.1), with a modified setting, in which the information content of the context is reduced because each target location was associated with 4 configurations (Exp.2).

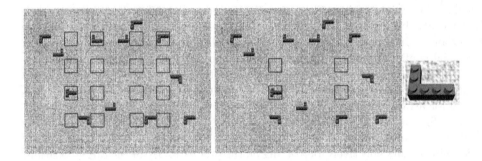

Fig. 1. Examples of search arrays and target locations of Exp.1 (left) and Exp.2 (right). Each object was made of 2 LEGO® bricks (small picture). The target was a rotated T presented at one of the marked locations (not marked in the actual experiment), the distractors were rotated Ls.

3 Hypotheses

In our first experiment, we used 16 target locations (cf. Fig. 1, left). These 16 locations were used in both, OLD and NEW configurations to avoid a reduction in probability just by the distinction NEW vs. OLD. At the beginning of the experiment, the probability to find the target by chance was 1:16. As in previous experiments, an association between OLD configurations, i.e. the arrangement of the items, and the target locations should be learned, resulting in an additional decrease in search times in the time course of the experiment (contextual cueing effect).

In a second experiment, the number of target locations was reduced to 4 (cf. Fig. 1, right) but the number of elements within each configuration and the number of configuration were the same. Again, in each trial of the experiment, possible target locations had to be searched to find the target. The probability to find the target by chance decreased to 1:4 for both OLD and NEW configurations. That means, finding a target should be faster in general in comparison to 16 locations because fewer locations have to be scanned.

The probability to find a target among similar distractors at a specific location depends, amongst others, on the number of possible target locations. If a search displays is repeatedly presented (OLD configuration), an association between context and target location is learned. The association reduces the number of locations that have to be scanned to find the target. The larger the number of possible target locations the larger the reduction of to-be-scanned locations. This property of the context can be described as the *predictive value* (as part of the *information content*) of the context. Is the context still used to guide attention to the target location if the information content and thus the predictive value of the context is low? And if so, how large is the contextual cueing effect and how is the time course of learning? In order to examine how contextual cueing is established we analyzed the time course of the effect, i.e. the difference in RT

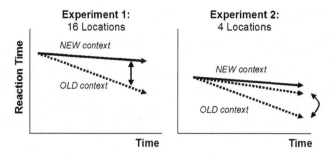

Fig. 2. Predictions of the contextual cueing effect, i.e. the RT differences between OLD (dotted lines) and NEW (solid lines) configurations, when the number of target locations varies (left Exp.1 and right Exp.2)

between OLD and NEW context was analyzed as a function of time, while the number of possible target locations varied (Exp.1 vs. Exp.2).

According to our assumption, we can make two predictions (cf. Fig. 2): 1) Reaction times should be lower when the number of possible target locations is reduced. 2) If contextual cueing depends on the information content of the context, the contextual cueing effect is diminished when the number of possible target locations is reduced. That is, the RT advantage of target localization in OLD contexts compared to NEW contexts caused by attentional guidance should be smaller.

4   Methods

4.1   Experimental Design

We ran two experiments, which were identical in the used stimuli and the general procedure. In both experiments, subjects were presented with 16 OLD and 16 NEW configurations in each of 30 blocks. The only difference between the experiments was the number of possible target locations. While in Experiment 1 a target could appear at one of 16 locations (each target location was used once for OLD and once for NEW configurations), in Experiment 2 a target was presented at one of 4 locations (each target location was used four times for OLD and four times for NEW configurations) (cf. Table 1).

Table 1. Number of configurations and target locations in each experiment

	Number of		Number of	Ratio
	OLD Configurations	NEW Configurations	Target Locations	Configurations to Locations
Experiment 1	16	16	16	1:1
Experiment 2	16	16	4	4:1

4.2   Participants

Thirty-two paid volunteers participated in two experiments, 16 participants each (14 female, 2 male, aged between 19 and 29 years, mean age Exp.1: 22.6 years; Exp.2: 22.7 years). In Experiment 1, three participants were left handed; in Experiment 2 all participants were right handed. All participants reported normal or corrected-to-normal visual acuity. The study was carried out in accordance with the ethical standards laid down in the 1964 Declaration of Helsinki.

4.3   Stimuli and Apparatus

Participants were seated in a comfortable armchair in a dimly lit and sound attenuated room with response buttons located under their left and right index fingers. All stimuli were presented on a 17 inch computer screen placed 100 cm in front of the participants at the center of their field of vision.

Participants searched for one of two task-relevant objects (the target) among other task-irrelevant objects (the distractors) and identified the target by pressing a response key. Each search array consisted of 12 LEGO® objects (1.2° in visual angle), which could appear within an invisible matrix of 12 x 9 locations that subtended approximately 15.1° x 10.2° in visual angle. The objects were designed by a CAD program (LegoCad). The software, which was developed by Lego in cooperation with Autodesk, allows constructing simple 3D models and machines. Two red LEGO® bricks were used to build one object. To make the stimulus set more realistic the objects were additionally rotated in perspective so that they appeared to lie on a surface.

In both experiments, the target was a LEGO® object in the form of a 'T', rotated 90° to the right or left. In Experiment 1, the target was presented at one of 16 selected locations of the 12 x 9 matrix (Fig. 1, left). In Experiment 2, the target was presented at one of 4 selected locations (Fig. 1, right). Thus, in Experiment 1 each target location was associated with exactly one repeated display, whereas in Experiment 2 each target location was associated with 4 different repeated displays (cf. Table 1). Target positions were identical for all participants of an experiment.

The distractor objects were 11 L shaped LEGO® objects, which were presented randomly in one of four rotations (0°, 90°, 180°, 270°). The distractor locations in each configuration were randomly sampled from all 108 possible locations, including target locations used in other configurations. In each configuration, half of the objects were placed left and the other half right of fixation, balanced for eccentricity. Configurations were generated separately for each participant.

Similar to previous experiments, we defined the visual context as the arrangement of distractor objects. The OLD set of stimuli consisted of 16 configurations, randomly generated at the beginning of the experiment then repeated throughout the entire experimental session once per block. The target (left- or rightwards oriented) always appeared in the same location within any particular configuration and the identities of the distractors within their respective spatial locations were preserved. The target type (left- or rightwards pointing T) was randomly chosen so that the identity of the target did not correlate with any of the configurations. In contrast, in the NEW set of stimuli configurations of distracting context elements were generated randomly on each trial. Any differences in RT between OLD and NEW configurations can be interpreted as contextual cueing.

4.4   Procedure

The whole experiment consisted of three parts: training at the beginning of the experiment, followed by the actual experiment, and a recognition test at the end.

Participants were instructed to search for a rotated T and press one of two buttons corresponding to whether the top of the T was pointed to the right or to the left as soon as they could. They performed three training blocks of 32 trials each. A trial started with a fixation cross appearing in the middle of the screen for 500 ms. Afterwards, the search display was presented for 500 ms, and participants pressed a key to indicate the identity of the target (a left- or rightwards pointing T). After a brief pause of 1-2 s, the following trial was initiated by the computer. The training was necessary to familiarize participants with the experimental task and procedure and to minimize inter-subject variability.

The experimental session consisted of 30 blocks of 32 trials each (16 OLD, 16 NEW configurations), for a total of 960 trials for each participant. Stimulus presentation and participants' task were identical to the training session. Feedback was given at the end of the block on the percentage of correct responses. Participants were not informed that the spatial configuration of the stimuli in some trials would be repeated, nor were they told to attend to or to encode the global array. They were simply given instruction to respond to the target's identity. It was stressed that they were to respond as quickly and as accurately as possible. A mandatory break of about 1-2 minutes was given after five blocks each and a longer break was given after half of the experiment.

At the end of the final block, participants performed a recognition task. The recognition served as a control measure. All 16 OLD configurations were presented again, intermingled with 16 NEWly generated configurations. Participants were asked to classify all configurations as already seen or new, respectively. If learning was indeed implicit, participants should not be able to distinguish between OLD and NEW displays.

4.5   Data Analysis

Reaction times were measured as the time between onset of the search display and the participant's response. Pressing the wrong button, pressing the button too quickly (<150 ms) and pressing it too slowly (>2000 ms) were defined as errors. Only correct responses were entered into statistical analyses. To estimate the general learning effect and the time point when context learning occurred, blocks were grouped in sets of 6 blocks each into 5 epochs. Error percentages and mean reaction times of both experiments were entered in repeated-measures ANOVAs with factors of context (OLD vs. NEW configurations), and epoch

(1 to 5) as within-subject factors and experiment (1 vs. 2) as between-subject factor.

Data were analyzed according to effects across experiments (common effects) and to effects between experiments (differences). A main effect of epoch would reflect changing RTs (or errors) in the time course of the experiment (i.e., general learning effect). A statistical main effect of context would reflect how repeating the same context affected the search for the target object (i.e., contextual learning). More important, if the difference between OLD and NEW context (i.e., the contextual cueing effect) varies over time (i.e., epochs), this would be reflected by an interaction between both factors and that would suggest the context information was learned over time (i.e., time course of the contextual cueing effect).

The main goal of the present experiments was to investigate the size of the contextual cueing effect when the advantage of context was reduced. Thus, in addition to the overall size, we also analyzed the time course of the contextual cueing effect, namely depending on the number of possible target locations. Statistically, this effect would be reflected by a three-way interaction between context (OLD vs. NEW), epoch (1 to 5), and experiment (1 vs. 2).

In order to demonstrate that the knowledge of display repetition (i.e., the context) is indeed implicit, a recognition test was performed at the end of the experiment. The hit rate (OLD displays were correctly categorized as old) was compared to the false alarms rate (NEW displays were wrongly categorized as old) by a paired t-test.

5   Results

Error percentages and reaction time data are presented as a function of epoch and context in Figure 3, separately for Experiment 1 and 2.

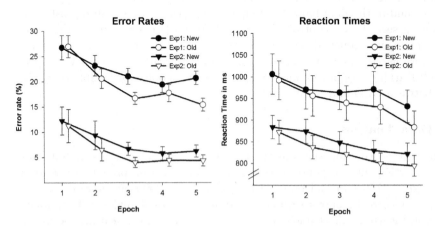

Fig. 3. Error rates (left panel) and reaction times (right panel) as a function of epoch (x-axis), separately for Exp.1 (circles) and Exp.2 (triangles)

In both experiments, participants' accuracy increased in the course of the experiment, $F(4,120)=19.7$, $p<.001$, $\epsilon=.142$, similarly, $F(4,120)<1$, reflected by decreasing errors from, on average, about 11.8% in epoch 1 to 5.3% in epoch 5. Planned comparisons revealed a significant decrease from epoch 1 to epoch 2, $F(1,30)=34.5$, $p<.001$, and from epoch 2 to epoch 3, $F(1,30)=7.2$, $p<.05$; error rates then remained constant on this level, all $F(1,30)<1$. Error rates also varied as a function of context, $F(1,30)=24.0$, $p<.001$; searching for a target in an OLD context was, on average, more accurate than searching in a new context (15.2% vs. 12.8%). However, a significant interaction between epoch and context, $F(4,120)=2.8$, $p<.05$, indicates that this advantage developed during the course of the experiment, indicating that an OLD context had to be learned as old first. Probably, only some of the repeated OLD displays were learned from block to block as reflected by a reliable linear trend, $F(1,30)=4.2$, $p<.05$. These effects were similar in both experiments since there was no interaction with the factor experiment (Fig. 3, left panel).

Table 2. Statistical Effects of Experiment 1 and 2

Statistical Effect	Description	Significance Error Rate	RT
Epoch	*General learning effect (GLE)*	$p<.001$	$p<.001$
Context	*Contextual cueing effect (CCE)*	$p<.001$	$p<.001$
Epoch x Context	*Time course of contextual cueing effect*	$p<.05$	$p<.01$
Experiment	*General differences between experiments*	$p<.001$	$p<.05$
Epoch x Experiment	*Differences in GLE between experiments*	*n.s.*	*n.s.*
Context x Experiment	*Differences in CCE between experiments*	*n.s.*	*n.s.*
Epoch x Context x Experiment	*Differences in time course of the CCE between experiments*	*n.s.*	$p=.05$

n.s. (statistically not significant): $p>.296$

Similarly to error rates, RTs decreased over time, $F(4,120)=11.9$, $p<.001$, $\epsilon=.580$ (Fig. 3, right panel). Single planned comparison showed a reliable linear trend in search time, $F(1,30)=20.1$, $p<.001$. The contextual cueing effect, defined as an RT benefit in the OLD condition compared to the NEW condition across all epochs, was significant, $F(1,30)=43.3$, $p<.001$. The significant interaction between epoch and context, $F(4,120)=3.6$, $p<.01$, indicated that performance was similar for both context types at the beginning of the experiment but that learning of the OLD context led to faster reaction times for the latter when compared to NEW contexts. The overall contextual cueing benefit, measured as the difference between OLD and NEW configurations across the last three epochs [9] was 38 ms ($SD=29$ ms) in Experiment 1 and 28 ms ($SD=22$ ms) in Experiment 2 (Fig. 3, right panel).

The comparison between both experiments shows, that with about 20.9%, error rate was relatively high in Experiment 1 and significantly higher than in

Experiment 2, $F(1,30)=41.1$, $p<.001$, where errors occurred on about 7.1% of the trials. Participants were also faster in finding the target (RT $\pm$ SD) when it was presented at one of four possible locations (Exp.2: 838 $\pm$ 93 ms) than when presented at one of 16 locations (Exp.1: 954 $\pm$ 160 ms). More interestingly, although the overall contextual cueing effect did not differ between both experiments, $F(1,30)<1$, learning was different, $F(4,120)=2.4$, $p=.05$ (cf., Fig. 4). In Experiment 1, the contextual cueing effect can be best described as linear trend, $F(1,15)=13.5$, $p<.01$, whereas in Experiment 2 planned comparisons of subsequent epochs showed that an improvement in contextual cueing only occurred from epoch 1 to 2, $F(1,15)=4.6$, $p<.05$, and remained constant at this level, all $F(1,15)<1$.

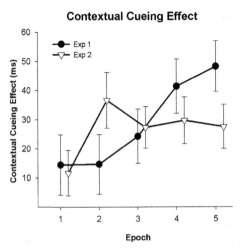

Fig. 4. Contextual cueing (RT differences between OLD and NEW configurations) as a function of time (x-axis), separately for Exp.1 (circles) and Exp.2 (triangles)

Similarly to previous studies, the knowledge about repeated configurations was implicit. In the recognition test, the hit rate (50.5% or 45.0%, respectively) did not differ significantly from the false alarms rate (44.1% or 48.4%, respectively), $F(1,30)<1$.

To summarize the results, we found contextual cueing effects of comparable size in both experiments. However, the time course of contextual learning differed, as clearly visible in Figure 4.

6   Discussion

The aim of the present experiments was to investigate how the information content of the visual context influences the contextual cueing effect. When the number of possible target location is reduced, the probability to find the target by chance is enhanced and the context provides less information in order to find the target. Thus, one might predict that the contextual cueing effect disappears or is at least reduced.

In both of our experiments, we observed a reliable contextual cueing effect: Finding a target in an OLD (familiar) context was faster than finding the target in a NEW context. Because the context (i.e., the spatial arrangement of the distractor items in a particular configuration) was univocally associated with the target location in that particular configuration, observers obviously used the context to guide attention to the relevant target location. Alternatively, one may argue that the repetition of some displays only increased sensitivity. This account would predict facilitation from repetition, which primes early perceptual processing mechanisms [9]. However, recently an electrophysiological study showed that early visual processes were not influenced by the familiarity the visual context [7]. Additionally, Wolfe and colleagues [8] found that the search for the target was not facilitated when a repeated context did not predict the target location.

The more plausible interpretation is that the visual context is used to guide visual attention to the target location. In the OLD condition, the arrangement of the distractors, the context, and the target location were kept constant. The target identity, however, differed. To find and identify the target, observers had to allocate their attention to the target location. In contrast to the NEW context condition, the context predicted the target location (but did not predict the target identity). Faster RTs in the OLD context therefore reflect a faster shift of attention or, in other words, the visual context guides visual-spatial attention. This interpretation is supported by electrophysiological experiments [7, 10].

In the present experiments, contextual cueing occured independently of the number of target locations. The size of the contextual cueing effect (as measured as the difference between OLD and NEW displays in the last three epochs) was similar. However, how the association between a specific repeated configuration and the corresponding target location was learned differed. This is clearly visible when the contextual cueing effects (i.e., the difference in RT between OLD and NEW displays) of both experiments are plotted against time (cf. Figure 4). When a target was presented at one of 16 possible target locations, contextual learning was linearly increasing with each repetition (Exp. 1). When a target could occur only at one of four possible locations, context learning was very fast (Exp. 2). It reached its maximum already in epoch 2 (after 6 repetitions).

7   Hypotheses of Attention Shifts in Contextual Cueing

How can these effects be explained? Most researchers agree that an association between the spatial arrangement of the distractors (or context) and the target location is formed, which is used to guide visual-spatial attention to the target location [e.g., 1 but see also 11]. There are (at least) two hypotheses how this association might be learned and used afterwards. In their initial paper, Chun and Jiang [1] proposed that contextual information is instance-based, and contextual cueing is a form of memory-based automaticity. These theories [12] assume that performance improvement is based on retrieval of past solutions from instances of past interactions stored in memory. When searching for a target, its detection is first mediated by generic attentional mechanisms in early stages of training. Continuing the task, memory traces of these interactions are established. These accumulate to provide solutions to the search task more quickly than a memory-free attentional mechanism would. It is assumed that these memory traces are instance-based, allowing for a distinction between stimuli that were presented in the history of perceptual interactions from novel stimuli that were not. According to these theories, the exact target location should be associated with a particular configuration. This hypothesis would predict increasing contextual cueing effects in the time course of the experiment until all repeated displays are learned. When an OLD display is presented again, attention can be shifted to the exact target location as the complete association is available. We call this *'instance-learning hypothesis'* in the current paper.

Alternatively, it would also be possible that not the exact target location is associated with a specific configuration but, as a first estimation, attention is roughly shifted close to the actual target location. For example, participants might first learn whether a target is located left or right of fixation, next the quadrant might be estimated before finally the exact target location may be associated with the context configuration. Thus, during the sequence of the experiment, rough estimations might first be computed for all OLD displays in a block and improve whenever these displays are repeated (*'estimation hypothesis'*). Presenting an OLD display again decreases RT because the area in which one has to search for the target becomes smaller.

In both cases, the learned context is used to guide visual-spatial attention. According to the 'instance-learning hypothesis', attention is guided directly to the exact location of the target but only for a learned context; the number of learned contexts increases over time. In the case of the 'estimation hypothesis', however, attention is guided roughly to the target area at the beginning of the experiment and more exactly with each repetition, resulting in decreasing search times because the area to be inspected becomes smaller.

According to the instance-learning hypothesis, a context-target association has to be formed for each of the 16 OLD contexts. In both experiments, each of the 16 OLD configurations has one unique target location that has to be associated with a particular context. As only some of these associations are learned during blockwise repetitions, the contextual cueing effect would increase until all of them are learned. Thus, learning should be independent of the number of possible target locations but depend on the number of associations to be learned (which depend on the number of OLD contexts). As the number of OLD contexts were the same in both experiments, the time course should also be the same.

The estimation hypothesis, however, predicts different time courses of the contextual cueing effect for both experiments. In case of 16 target locations (Exp.1), the area to be searched for the target is reduced only slowly in the time course of the experiment. Because of the high number of possible target locations, it takes some time until attention is shifted precisely to the correct location. Thus, the prediction would be very similar to the instance-learning hypothesis (Fig. 5, left panel). If, however, a target can only appear at one of 4 possible locations (Exp.2), already a rough estimation of the target location is very efficient in guiding attention to the target. Thus, learning should be fast at the beginning of the experiment, reaching an asymptote very quickly (Fig. 5, right panel). Comparing the time course of contextual cueing of both experiments, our data support the estimation hypothesis.

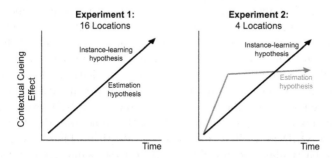

Fig. 5. Predictions of both hypotheses. When a target appears at one of 16 possible locations, both hypotheses would predict the same time course of the contextual cueing effect (left panel). If, however, a target appears only at one of four locations, the instance-learning hypothesis would predict the same time course as before while the estimation hypothesis would predict fast learning at the beginning, reaching an asymptote soon (right panel).

Because the time course of contextual cueing differed between both experiments and the time course of the second experiment followed the predictions of the estimation hypothesis, we assume that attention is probably not shifted directly to the exact stimulus location but the number of probable target locations is minimized by a rough estimation. Going back to the example described at the beginning, this would mean: We do not need to know the exact location of our cup on the desk to find it efficiently in a familiar context. Rather, we find it faster when we only roughly know the part of the desk the cup is placed in.

This idea can be further investigated by the so-called eye tracking method, by which participant's eye movements are measured while they are performing a contextual cueing task. It is well established in literature that one can estimate the allocation of attention by measuring where people look at. Using this method in a contextual cueing task allows to determine where attention has been (fixation location). Also, the accuracy (and inaccuracy) of attentional guidance (the distance of a fixation from the target) can be measured, as well as the number of items that were attended before the target was found (the number of fixations). A recent study by Peterson and Kramer [13] partly supports the assumption that attention is not directly guided to the target location. In one of their experiments they showed that only on 11.3% of the trials, the eyes (and therefore also attention) went immediately to the target. For comparison, the chance probability that the first item inspected by the eyes was the target was 8.33%. However, the first fixation was not closer to the target in OLD than in NEW configurations, a result that seems to contradict our interpretation of the results. The authors argue that the brain does not notice a repeated display from the beginning of its presentation but that recognition occurs later in the search process. As soon as the context is recognized, attention is guided directly to the target location.

In their model, recognition of the repeated search display plays an important role. The authors assume that the search for the target is started in a similar way in OLD and NEW configuration. As soon as some parts (one or more objects and their spatial relations to each other) of the contexts are recognized, the associated target location is retrieved from memory and applyed to the current search display in order to guide attention. It should be noted that the term 'recognition' does not imply that participants can explicitly remember a specific context. This model does explain the occurence of a contextual cueing effect in our experiments but not the different time courses. In both experiments, we used the same number of search displays with the same number of items. Also, the number of OLD and NEW displays was identical. Thus, the probability to recognize a particular configuration was exactly the same. However, although the size of the contextual cueing effect was comparable between both experiment (that would be expected by Peterson's and Kramer's interpretation), the time course across the experiment varied (a result that cannot be explained by their findings).[1]

8   Application to Technical Systems

Within the engineering domain, the development of enabling technologies such as autonomous robotic systems and ambient intelligence systems involves the real-time analysis of enormous quantities of data. These data have to be processed in an intelligent way to provide 'on time delivery' of the required relevant information. Knowledge has to be applied about what needs to be attended to, and when, and what to do in a meaningful sequence, in correspondence with visual feedback. Contextual cueing may be one important process that allows shortening the computational time required to estimate where attention should be directed to.

Depending on the proposed hypotheses, the algorithm to be implemented would differ. The instance-based hypothesis would require instances for all possible contexts. Although very effective if a context is once learned, a slightly different context would require learning of an additional instance. If, however, an estimation algorithm would be implemented, the learned knowledge of the

---

[1] However, we cannot rule out the possibility that in our experiments the learning process itself (i.e., how the association between context and target location was acquired) differed when the number of target locations was reduced. This alternative interpretation of the present results has to be tested further, for example, by measuring eye movement or the underlying brain activity.

system could also be transferred to similar contexts and a target could still be found efficiently when it is placed slightly beside the original target location.

Recently, some models have been developed to implement contextual knowledge into attention models in order to predict its guidance [14, 15, 16]. Because of limited space we are not able to describe them in detail but it should be noted that visual context plays an important role in guiding visual-spatial attention.

To summarize, although visual search for a task-relevant object among task-irrelevant objects is a demanding task for the visual information processing, implicit learning mechanisms, such as contextual cueing, allow the visual system to quickly extract stimulus regularities [2]. Implementing such a mechanism into cognitive technical systems may help to develop flexible and adaptive behavior. The current experiments provide some evidence how such a learning process might be implemented to technical systems. A rough estimation algorithm where to shift attention to might be the better way to simulate context information than learning of specific context-target associations.

9   Conclusion

When search displays are presented repeatedly, an association between the context (i.e., the configuration of distractors) and the target location is formed, which guides attention and speeds up search time. Depending on the number of locations, the time course of the contextual effect can differ even when the averaged effect is similar. Thus, the time course of this effect provides important information about the underlying contextual learning.

Acknowledgments

This project was part of the Cluster of Excellence 'Cognition for Technical Systems' (CoTeSys, Project 148) funded by the Deutsche Forschungsgemeinschaft (German Research Foundation, DFG). We want to thank Olaf Stursberg for helpful comments on an earlier version of this manuscript and Christian Stößel for his help in designing the stimuli.

References

[1]   Chun, M.M., Jiang, Y.H.: Contextual cueing: Implicit learning and memory of visual context guides spatial attention. Cognitive Psychology 36, 28–71 (1998)

[2]   Chun, M.M., Jiang, Y.H.: Top-down attentional guidance based on implicit learning of visual covariation. Psychological Science 10, 360–365 (1999)

[3]   Chun, M.M., Nakayama, K.: On the functional role of implicit visual memory for the adaptive deployment of attention across scenes. Visual Cognition 7, 65–81 (2000)

[4]   Jiang, Y.H., Chun, M.M.: Selective attention modulates implicit learning. Quarterly Journal of Experimental Psychology Section a-Human Experimental Psychology 54, 1105–1124 (2001)

[5]   Endo, N., Takeda, Y.: Selective learning of spatial configuration and object iden-
      tity in visual search. Perception & Psychophysics 66, 293–302 (2004)
[6]   Kunar, M.A., Flusberg, S.J., Wolfe, J.M.: Contextual cuing by global features.
      Perception & Psychophysics 68, 1204–1216 (2006)
[7]   Schankin, A., Schubö, A.: Cognitive processes facilitated by contextual cueing.
      Evidence from event-related brain potentials. Psychophysiology (in press)
[8]   Wolfe, J.M., Klempen, N., Dahlen, K.: Postattentive Vision. Journal of Experi-
      mental Psychology: Human Perception & Performance 26, 693–716 (2000)
[9]   Bar, M., Biederman, I.: Subliminal visual priming. Psychological Science 9, 464–
      469 (1998)
[10] Olson, I.R., Chun, M.M., Allison, T.: Contextual guidance of attention: human
      intracranial event-related potential evidence for feedback modulation in anatomi-
      cally early temporally late stages of visual processing. Brain 124, 1417–1425 (2001)
[11] Kunar, M.A., et al.: Does contextual cuing guide the deployment of attention?
      Journal of Experimental Psychology: Human Perception & Performance 33, 816–
      828 (2007)
[12] Logan, G.D.: Towards an instance theory of automatization. Psychological Re-
      view 95, 492–527 (1988)
[13] Peterson, M.S., Kramer, A.F.: Attentional guidance of the eyes by contextual
      information and abrupt onsets. Perception & Psychophysics 63, 1239–1249 (2001)
[14] Brady, T.F., Chun, M.M.: Spatial constraints on learning in visual search: Mod-
      eling contextual cueing. Journal of Experimental Psychology-Human Perception
      and Performance 33, 798–815 (2007)
[15] Backhaus, A., et al.: Contextual learning in the selective attention for identi-
      fication model (CL-SAIM): Modeling contextual cueing in visual search tasks.
      In: Proceedings of the IEEE-CVPR Workshop in Attention and Performance in
      Computer Vision (WAPCV), pp. 1–7. IEEE-Press, San Diego (2005)
[16] Torralba, A., et al.: Contextual guidance of eye movements and attention in real-
      world scences: The role of global features in object search. Psychological Re-
      view 113, 766–786 (2006)

Conspicuity and Congruity in Change Detection

Jean Underwood[1], Emma Templeman[2], and Geoffrey Underwood[3]

[1] Division of Psychology, Nottingham Trent University, Nottingham NG1 4BU, UK
jean.underwood@ntu.ac.uk
[2] School of Psychology, University of Nottingham, Nottingham NG7 2RD, UK
lpyyec1t@nottingham.ac.uk
[3] School of Psychology, University of Nottingham, Nottingham NG7 2RD, UK
geoff.underwood@nottingham.ac.uk

Abstract. How does visual saliency determine the attention given to objects in a scene, and is the detection of change dependent upon the conspicuity of the changed object? Viewers' eye movements were recorded during the inspection of pictures of natural scenes. Two versions of a scene were compared to determine whether or not they were the same. The two images were either available at the same time (Experiment 1), or consecutively (Experiment 2). When an object was changed, it either had high or low visual saliency and it either was congruent with the scene or it violated the gist in that it would not be expected to be seen in that context. Previous studies have indicated that incongruous objects sometimes attract early attention, but the inconsistency of this effect leads to the question of whether it is dependent upon conspicuity rather than congruity. Incongruous objects attract early eye fixations here, dismissing the explanation based on visual saliency.

1 Introduction

When inspecting a scene, incongruous objects are identified with greater difficulty than corresponding objects that are consistent with the gist. In each scene with a recognizable gist – a bathroom, a roadway, or a golf course, for example - we would expect to see certain objects – a sink, a bar of soap, a towel, and a toothbrush in the case of a bathroom. If an object from another scene is present – a golf ball in the sink, perhaps – it can be described as violating the gist. Such an object would be recognised with greater difficulty than if it had appeared in a picture of a putting green.

The conclusion that we have rapid identification of scene gist follows from Biederman's studies in which viewers attempted to identify objects in briefly presented photographs that were shown individually [1, 2, 3]. When pictures were cut up and rearranged, thereby disturbing the gist of the scene, objects that remained in their unjumbled locations were recognised with more difficulty. These experiments support the conclusion that the gist of the scene aids object identification, reflecting the facilitating effect of sentential context on word recognition when reading. The interaction between objects and their context of presentation has also been demonstrated by Davenport and Potter [4]. Objects were copied onto background scenes that indicated a contextual gist, and objects that violated the gist were identified less well than those

L. Paletta and J.K. Tsotsos (Eds.): WAPCV 2008, LNAI 5395, pp. 85–97, 2009.
© Springer-Verlag Berlin Heidelberg 2009

that were consistent (for example, a football player superimposed into the foreground of a church interior compared to a sports field).

The evidence from these studies suggests that the gist of a picture can be perceived quickly, and that context aids the recognition of objects that are consistent with the scene. Violations of gist result in impaired recognition, as they did with the gist-reduced jumbled pictures used by Biederman and his colleagues [1, 2, 3] and with the identification of inconsistent objects pasted into Davenport and Potter's [4] scenes. Objects can violate the gist by the improbability of their appearance in that scene. Biederman, Mezzanotte and Rabinowitz [5] found that objects in drawings had impaired identification when they were unlikely elements of that scene as well as when they violated the relational structure of the scene, for example by being drawn too small or by not resting on a supporting surface. This experiment again demonstrates that objects that violate the gist are recognised with greater difficulty than those that contribute to the gist.

This conclusion, which appears from the Biederman and Davenport studies [1, 2, 3, 4, 5], conflicts with the findings from studies of picture perception in which eye movements are recorded. Objects that violate the gist of the scene attract early eye fixations. This is paradoxical because some studies have demonstrated a recognition disadvantage while other studies have demonstrated that these objects are able to attract earlier eye fixations. Mackworth and Morandi [6] recorded eye movements while viewers judged which of two pictures they preferred. Regions of the pictures that were regarded subjectively as being most informative received more fixations, and non-informative regions were often not fixated at all. This suggests that our viewing of a picture can be guided by the processing of meaningful elements prior to their fixation and close inspection, and this conclusion was supported by Loftus and Mackworth's [7] experiment with line drawings of simple scenes in which an incongruous object was fixated very early in the sequence of inspection. Viewers tended to fixate the anomalous object with the first fixation, and they tended to fixate these objects earlier than a gist-consistent object drawn in the same region of the picture.

It is important to note that the idea of early fixations being attracted to anomalous objects has not gone unchallenged. Two studies have recorded eye fixations during the inspection of line drawings of familiar scenes that sometimes contained objects that were out of place, and incongruous objects attracted early fixations in neither experiment [8, 9]. In a range of tasks there was no evidence of the earlier fixation of anomalous relative to gist-consistent objects. One possibility that might explain this inconsistency in the pattern of results follows from differences in the images used in the different experiments. The stimuli in these later studies [8, 9] were redrawn from photographs and were therefore visually complex with crowded and partially occluded objects in them, whereas Loftus and Mackworth [7] used simple hand-drawn sketches that contained sufficient information to convey the intended gist but little detail. Identifying any object in the complex line-drawings is difficult but with the simple drawings each object is readily identifiable. The possibility to be considered here is that the early capture of attention by incongruous objects depends upon their visually conspicuity.

Attention is attracted by the conspicuity of objects and the early allocation of visual attention to an object is determined by its low-level visual saliency value relative to that of other objects in the scene. During the initial viewing of an image a saliency map is said to be developed using low-level visual discontinuities of color, intensity and line orientation [10]. The saliency peaks in this map represent regions that are distinct from their surroundings, and attention is first attracted to the highest peak – the most conspicuous object or region. The semantics of the scene can be appreciated only after early attention has been allocated to regions according to their saliency ranks. After the first few fixations the meaning of the scene can be appreciated, and eye movements can then be made to areas of high-level interest. The model makes good predictions about the early locations of eye fixations on static and dynamic pictures in free-viewing and recognition memory tasks [11, 12, 13, 14]. A plausible explanation of the effect in the Loftus and Mackworth [7] is that the anomalous objects in the pictures may have been more visually salient than those used in the later studies [8, 9], and it may have been high saliency that resulted in the attraction of early fixations. Saliency does not explain why Loftus and Mackworth found a difference between consistent and inconsistent objects, however, only why the difference does not emerge when the objects are obscured by rich backgrounds. Our initial purpose was to determine whether incongruous objects attract early eye fixations in pictures of real-world scenes when the saliency values of the objects are known and controlled. Photographs rather than line-drawings were used, and the saliency values of objects used to determine whether early eye fixations are associated with conspicuity or congruency.

In the first experiment eye fixations were recorded while participants looked at pairs of pictures in a comparative visual search task [15, 16, 17]. The two pictures were displayed side by side and the task was to say whether they were the same or not. On those trials where the two pictures were not identical, only one object was changed. The scenes had readily identifiable gists and were photographed in a domestic environment. Each scenes was photographed a second time with just one object replaced, and the new object was either consistent with the gist, or it was incongruous in that it would not normally be found in that setting. The saliency values of all objects were determined used the Itti and Koch algorithm [10], and a set of pictures created in which the new object varied in its low-level conspicuity and in its high-level congruency.

In the second experiment the same images were used and eye fixations again recorded, but here the two pictures were presented sequentially. The comparative visual search task was used because it requires a search of objects in the scene rather than a search for one object. Viewers characteristically make a series of comparative brief fixations, looking first at an object in one picture and than at the corresponding region in the other picture to determine whether they are identical or not [17]. Detection of a difference depended upon direct fixation of the changed object in that study, and so we expected a high proportion of fixated target objects here. By presenting the second image only after the first image has been inspected and its display terminated, this task emphasizes the role of visual memory in change detection, whereas in the comparative visual search task the viewer need only remember one object at a time before switching fixation to the corresponding region of the paired image. The successive-presentation version of the task requires encoding of an entire scene rather than of individual objects within that scene.

2   Experiment 1: Comparing Concurrent Images

In this experiment viewers looked at a pairs of images to decide whether they were identical or whether an object had been changed. Both versions of the pair of images remained on the screen until a response was made. Eye movements were recorded to determine whether the early fixation of incongruous items were associated with their visual saliency.

2.1   Method

Twenty-four members of the university community who had normal or corrected-to-normal vision were paid for their participation in this experiment.

Pairs of digital color photographs were prepared for presentation on a 36 cm x 27 cm computer monitor, with 40 pairs of identical images and 40 pairs with one object replaced in otherwise identical images. For the pairs containing a change, two photographs were taken from the same position, with an object replaced by another of similar size. The scenes were photographed in portrait orientation, so that they could be displayed side by side on the monitor, and there was a gap of 0.5 cm (0.46 deg). Each of the paired images was shown at a viewing distance of 60 cm and subtended 17.5 deg x 22.5 deg at this distance.

For the pairs of pictures that contained a changed object, four types of changes were made. Changed objects could vary in their visual conspicuity as well as their semantic congruency. The high conspicuity condition was defined by one of the objects having high saliency. For the purposes of the experiment, high saliency was defined as being one of three top-ranked objects according to the Itti and Koch (2000) algorithm, and low saliency was not in the eight most salient objects. The two conditions that are marked as having conspicuous target objects have much lower mean ranks than do the two conditions marked as showing inconspicuous objects. The mean ranks for the conspicuous conditions were very similar (1.68 and 1.55 for congruent and incongruent changes), as were the mean ranks for the inconspicuous conditions (13.25 and 13.70). This close matching of the saliency ranks of the target objects ensures that any effects of congruency are independent and cannot be attributed to conspicuity. The original object was always congruent to the scene and of a saliency rank value of 4-7. The semantic congruency of a scene was either maintained by using objects normally associated with the other objects depicted (congruent), or violated by replacing it with an object from another scene (incongruent). In an image of a bathroom shower, for example, a food container replaced a shampoo bottle, and for a congruent change in a scene showing food preparation on a kitchen worktop a potato replaced an apple. There were ten examples of each type of change. An example of a pair of images with a change is shown in Fig. 1 (top pair), with eye fixations from one participant superimposed on them. Fig. 1 (bottom pair) also shows an example for a pair of identical pictures, again with the fixations of one participant indicated.

Forty additional photographs without any changes were prepared, with half containing an incongruous object. This enabled us to show participants equal numbers pairs without and without changes. Each of the 40 scenes was used in each of the four change conditions over the course of the experiment, with four images created of each scene. Each scene was presented in each condition to different participants, but each

Fig. 1. Examples of pairs of images in which there is a changed object (top) and in which the images are identical (bottom). Eye movements from one participant are represented by the super-imposed lines, and started in each case with fixation on the white space between the two versions of the image. Note the larger number of fixations when there is no object changed (bottom), and the sequence of comparative fixations from an object in one picture to the corresponding region in the other picture.

participant saw a scene only once. The images were displayed on a colour monitor and had a resolution of 1024 x 768 pixels. A head-mounted SMI EyeLink system was used to record eye movements, with recordings taken every 4 ms, with a spatial accuracy better than 0.5 deg. A chin rest was used to restrict head movements. The experiment started with a 9-point calibration procedure and once successful a set of 8 pairs of photographs was shown for practice and to demonstrate the type of pictures used. Examples from all conditions were used during practice. The participants were instructed to first fixate a marker in the center of the computer monitor, and when the pair of photographs appeared they were to say whether they were the same or different by pressing one of two keys on the computer keyboard. The pair of pictures remained on the screen until the participant responded.

2.2 Results

The keyboard response times were submitted to a 2x2 analysis of variance with congruency and conspicuity as the factors. The means and standard errors are shown in Fig. 2, together with the corresponding means from Experiment 2. The anova showed an effect of congruency of the replacement object (F(1,23) = 10.74, p < .01), with faster decisions to pairs of images containing incongruent objects (2.75 s) than to those with a congruous object (3.15 s). There was no main effect of conspicuity (F(1,23) = 1.57) and no interaction (F < 1). Response accuracy (see Fig. 3) was greater than 85% and there were no reliable effects of saliency (F<1) or congruency (F(1,23) = 3.62) and no interaction (F<1).

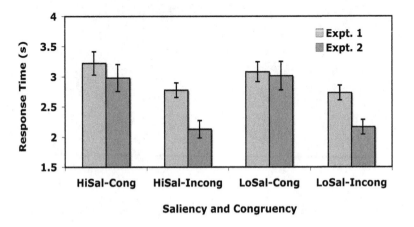

Fig. 2. Response times in Experiment 1 and Experiment 2, for the four saliency/congruency conditions. HiSal-Cong = high saliency and congruous; HiSal-Incong= high saliency and incongruous; LoSal-Cong = low saliency and congruous; LoSal-Incong = low saliency and incongruous.

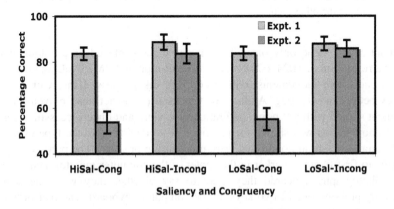

Fig. 3. Response accuracies in Experiment 1 and Experiment 2, for the four saliency/congruency conditions

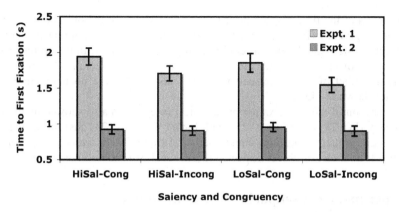

Fig. 4. Time elapsed prior to first fixation of the changed object in Experiment 1 and Experiment 2, for the four saliency/congruency conditions. Note that the large difference between experiments is due to the availability of two images in Experiment 1 and only one image in Experiment 2.

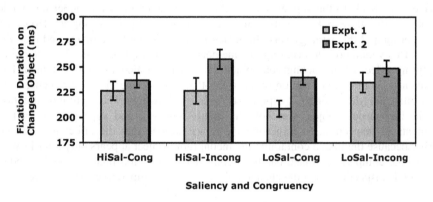

Fig. 5. Duration of the first fixation of the changed object in Experiment 1 and Experiment 2, for the four saliency/congruency conditions

The remaining analyses determined how quickly the objects of specific interest were inspected, using the time and number of fixations made prior to the first fixation on the changed object and on its counterpart original object in the pair of images displayed. In the pictures of a kitchen scene in Fig. 1 (top pair), the participant has made comparative fixations of the box of teabags in each image before moving to a nearby food jar. The comparative fixation reveals that the food jar in the picture on the left has been replaced in the picture on the right, by a shampoo bottle (incongruous changed object). The response decision can then be made. Three within-groups anovas, each with three factors, were used to analyze the number of fixations prior to fixation of each object, the time elapsed before fixation of each object, and the mean duration of the first fixation on each object. The three factors were object of interest (changed original), high/low congruency, and high/low saliency.

The anova of the time elapsed prior to the first fixation of the objects indicated an effect of congruency ($F(1,23) = 8.10$, $p < .01$), with pictures containing two congruous objects being fixated 1.90 s after display onset, in contrast with the corresponding incongruous object being fixated after 1.66 s. There was no effect of saliency ($F(1,23) = 2.21$) no effect of object inspected ($F < 1$), and there were no interactions. The means are shown in Fig. 4.

The duration of the first fixation on an object (see Fig. 5) also showed an effect of congruency ($F(1,23) = 4.57$, $p < .05$), with longer fixations (231 ms) on incongruous objects than on their congruous counterparts (218 ms). There was no effect of saliency, $F < 1$, no effect of object inspected ($F(1,23) = 1.38$) and no interactions.

3   Experiment 2: Comparing Consecutive Images

In Experiment 1 the participants looked at two co-present images to determine whether they were the same or not, and they frequently made comparative eye fixations on corresponding objects. This strategy is illustrated in Fig. 1, where successive fixations are made to an object in one image and then to the same region of the other image. This leads to the suggestion that they do not attempt to remember the whole scene, but that they make specific comparison between individual objects. However, in Experiment 2 this strategy was eliminated by showing first one image and then the other. This version of the task requires memory of the complete scene. The attention-capturing effect of an incongruous object was demonstrated to some extent in Experiment 1, with earlier fixation of an object that violated the gist of the scene, as in the Loftus and Mackworth [7] experiment with line drawings. The capture of attention by the incongruous object was not immediate, however, and appeared only a few moments of inspection. In Experiment 2 it was argued that the effect may appear earlier because the image containing the incongruous object appears only after inspection of the comparison image has been competed. If the first image is well-encoded during this inspection, then the changed object may be comparatively distinctive.

3.1   Method

Twenty-four participants with normal or corrected-to-normal vision were paid for their participation. None had been involved in Experiment 1.

The pictures from Experiment 1 were again used here, but whereas in Experiment 1 the two versions of an image appeared on the screen at the same time, in this experiment they appeared one after the other, on opposite sides of the computer screen. The first image of a pair appeared on the left and the right of the screen equally often. Viewing conditions were otherwise the same as in Experiment 1.

For each pair of different images, there were four variations of the changed object per scene, using the same four different treatment conditions of high/low saliency and congruent/incongruent. When a replacement object was used, it always appeared in the second image of each pair. As previously, each participant saw 40 pairs of images with an object changed, and 40 pairs where there was no difference between them. The eye-tracker and calibration procedure were as described in Experiment 1.

3.2 Results

The response time, accuracy, and eye fixation measures were subjected to within-groups anovas, with saliency (high vs. low) and congruency (congruent vs. incongruent) as the factors. Although eye movements were recorded during the inspection of both images in each pair, only the data from inspection of the second image were analysed.

The response times were reliably different in the congruency condition (F(1,23) = 35.91, p <.01) with faster responses to trials containing an incongruent object (2.15 s) than to trials containing a congruent object (2.99 s). There was no main effect of saliency (F<1) and no interaction (F<1).

Responses also varied in accuracy according to the congruency of changed objects (F(1,23) = 54.35, p <.01), with more accurate responses to pictures containing an incongruent object (84.79%) than to pictures containing a congruent object (54.38%). There was no main effect of saliency (F<1) and no interaction, F<1. This pattern is shown in Fig. 3.

The anova of the time elapsed prior to first fixation on a target object did not reveal a main effect of saliency (F<1), or of congruency (F<1), or an interaction, F<1.

The congruency of an object influenced the duration of the first fixation (F(1, 23) = 15.57, p <.001), with a longer fixation on an incongruent target object (254 ms) than a congruent target object (239 ms). There was no main effect of saliency (F<1) and no interaction (F<1) between the factors of congruency and saliency.

4 Discussion

In each experiment viewers saw two pictures of a natural scene on each trial, and judged whether they were the same or different. When they were different just one object had been changed. The new object was either high or low saliency, and was either semantically congruent or incongruent. In Experiment 1 the two versions of each picture were presented concurrently, and in Experiment 2 they appeared consecutively. To identify the changed object with the concurrent display (Experiment 1), the viewer could look at an object in one version of the picture and then check whether it was changed or unchanged in the other version. The iterative use of this object-by-object inspection strategy eventually finds the changed object, when there is one, and this strategy is commonly found in studies of comparative visual search [17]. With the consecutive display (Experiment 2) a different strategy is required, and the entire first image must be encoded prior to the second image becoming available. This task requires a visual memory of the image to be compared against the second version of the picture. This consecutive comparison task was difficult – accuracy was poorer, and fixation durations were longer than when object-by-object inspection was possible.

The other main difference between these two experiments involved the time that elapsed before the changed object was fixated. When both images were available an effect of semantic congruency was found – incongruent objects attracted attention and were fixated earlier than objects that did not violate the gist of the scene. This effect was not apparent with the consecutive displays when the whole scene had to be encoded and remembered. This task can be considered to be a one-shot version of the traditional change detection task with the flicker paradigm [18, 19]. When single objects are changed in natural scenes that repeatedly alternate between the two versions of the image [20], detection accuracy is similar to that seen in Experiment 1. The poor performance in the one-shot version of change detection used here is attributable to the need to remember the detail of an entire visual scene from a single viewing, and our reputably excellent visual memory does not serve this task well.

The study investigated whether incongruent objects attracted early fixations and whether this incongruency effect depended upon the object being visually conspicuous. The question about attentional capture by objects that violate the gist was prompted by an inconsistency in the results from previous studies of incongruity [7, 8, 9]. This question is important because it bears on the issue of whether objects in scenes can be recognised pre-attentively using peripheral vision. If pre-attentive recognition of objects in natural scenes is possible, then we might expect to see informative or interesting objects, such as those that violate the gist, being recognised early and without foveal inspection. When objects attract eye fixations they must have been processed to the extent that the differentiating feature can be used by the eye guidance mechanism – whether it is a low-level feature such as colour or a high-level feature such as semantic incongruity. If it is the scene semantics that influence guidance then we can conclude that the meaning of the object/scene relationship has been recognised prior to the object receiving focal attention.

Experiment 1 provides partial support for the view that objects can be recognised prior to the fixation and that this process of recognition can be used to guide subsequent eye movements. The evidence is supportive in that incongruent objects that violated the scene gist were fixated earlier than objects that were consistent with the gist. Incongruent objects were fixated earlier than their congruent counterparts. Two other measures also indicated an effect of object congruency: the longer inspection of the display, and longer durations of fixations on the object, when the display contained an incongruous object. The sensitivity to the congruency between the object and the scene indicates that the object is identified to some extent prior to its first fixation. This does not establish that an incongruent object is fully identified prior to its first fixation, only that some characteristic has been recognized, establishing that it is not semantically consistent with the other objects in the scene. Prior to their fixation then, objects may be partially recognised and their relationship with other objects determined to be inconsistent. This incomplete processing may involve the recognition of the shape, colour and other visual features that together will identify the object as belonging to a category of items that are improbable members of that scene. In the kitchen scene shown in Fig. 1 for example, a bottle of the particular shape is improbable in the location shown, and the viewer does not need to identify it as a shampoo bottle to appreciate the incongruency. Detailed foveal inspection provides this fine-grained information, but it may be the early partial identification that demands fixation of the object. When it is eventually inspected, the first fixation on the object is longer than if the object had been congruent with the context of the scene, and this long foveation was seen in both experiments.

The results provide only partial support for the early recognition of objects because the first fixation of the critical object did not occur until several fixations after first inspection of the pictures, that is the effect was not apparent until the pictures had been displayed for several seconds. Loftus and Mackworth [7] reported a difference in the probability of fixation between congruent and incongruent objects on the second fixation, suggesting that the incongruency was recognised almost immediately upon the appearance of the display. In Experiment 1 there was no evidence of such an early difference in the fixation of objects (see Fig. 4), and 1.66 s elapsed before fixation of the incongruous object. This was faster than the corresponding number of fixations prior to fixation of a congruous object (1.9 s), but not indicative of the immediate identification of an object that violated the gist of the scene. Unlike the earlier study [7], an incongruous object did not attract the first or second fixation on the picture. This may have resulted from our use of natural images that contained greater complexity than Loftus and Mackworth's line-drawings. More detailed inspection with more fixations may have been necessary in the present experiments, to build a usable representation of the image.

The inconsistency between results from previous studies that have used line-drawings [7, 8, 9] is that the incongruency effect may emerge only when the incongruent object is visually salient and is conspicuous against its background scene. This possibility was explored by varying the saliency of the object that was changed. Itti and Koch's [10] algorithm for determining visual saliency was applied to all of the images used, and the objects of interest had either high saliency (being one of the most conspicuous objects in the picture) or low saliency (being relatively inconspicuous). There was no effect of conspicuity according to any of the measures taken in either experiment. A similar result was observed in a change detection experiment using the flicker paradigm [20], with incongruous objects detected most easily, but with no effects of saliency. In the present study an incongruent object attracted attention early and gained faster responses than those where the gist was not violated, but there were no indications of an effect of saliency with these or with any of the other measures. The incongruency effect cannot depend upon the visual conspicuity of an object that violates the gist of the scene as it was not the brightness or the colour of an incongruent object that resulted in its early fixation. The absence of an effect of saliency demonstrates that the result is a consequence of the viewer's sensitivity to the scene semantics rather than from low-level visual processes.

One might argue the incongruity effect is artifact of the method, in that any incongruous items were always changed items, so identification of an incongruous item would lead to an immediate cessation of the search. However, this assumption is consistent with the response time data but cannot explain differences in fixation patterns, specifically the early fixation of incongruous objects

Loftus and Mackworth's (1978) three-stage model of scene perception is largely confirmed by the current evidence. These stages are the rapid determination of the gist of a scene, the partial recognition of objects in the scene, and the computation of the conditional probabilities of objects appearing in that scene. The eye guidance mechanism then makes use of these conditional probabilities, directing fixations to objects with low probability of occurrence. The first two stages must occur in parallel, otherwise we have the conundrum of the gist being recognised prior to identification of the component objects. If there is partial identification of the objects – we might recognise an object as a bottle, or a piece of fruit without identifying the type of bottle or the type of fruit – then the gist of the scene can be established as the sum of the features identified. An object that violates the gist may then cause a perturbation or discontinuity in this semantic saliency map and this activity will attract fixations as the viewer attempts to resolve the inconsistency. This model of scene perception may resolve the paradox of why incongruent objects are difficult to identify and yet attract eye fixations. The partial recognition of a non-fixated object may be sufficient to determine that it violates the gist of the scene and that it requires more detailed inspection, and when it is fixated the inconsistency between the object and its context delays full identification.

The difficulty of the task in Experiment 2, in which viewers' were required to remember the first scene for comparison with the second, supports Rensink's [21] coherence theory to some extent, in that it implies that no lasting memory of an object in a scene is formed once attention has left that object. If a stable visual memory of the first scene had been established the changes would have been detected more accurately. We clearly do have visual memories of natural scenes, and so the possibility arises that short-term visual memories impose a high cognitive load and that makes their use in change detection inefficient. Object changes can be detected after long intervals however – after a 30 min interval [21] and even after 24 hours [22]. Hollingworth [22] found that after a 20 s scene presentation, rotation changes or token changes of an object could still be detected above the level of chance, even though detection rates were impaired compared to an immediate test. It was concluded that this finding endorsed the robustness of visual memory theory of scene representation that suggests that when an object is fixated, an object representation is fixed to the spatial layout of the scene in short-term memory, which is then consolidated into a long-term memory representation [21]. According to this theory, when a target object was inspected, the object from the first picture that was encoded to that specific location in a spatial memory representation should have been activated, thus allowing the change to be detected. However, the accuracy data do not confirm this argument. The discrepancy in these results with those of Hollingworth [22] may be due to task differences. In this experiment, participants were told to search for a change whereas in Hollingworth's experiment, the target object was cued. Therefore, it may not the case that no scene representation indexed with objects and their location was formed after viewing the first picture in each pair, but that access to specific object representations requires a retrieval cue. While this suggests that a retrieval failure may be responsible for the inaccuracy of the results, it cannot explain the large difference in accuracy between the congruent and incongruent target objects and it therefore seems unlikely. An alternative explanation of the poor accuracy of responses in the consecutive display task may be that the original objects were not encoded as thoroughly as required by the task. Yet again, an encoding failure cannot explain why trials with incongruent target objects were responded to more accurately than those with congruent target objects.

In essence it appears from the accuracy data that visual memory representations of the first picture in each pair were either not created or just not used effectively. This complies with O'Regan's [24] suggestion that our environment is rich enough to serve as its own 'external memory', so when presented with a scene we do not need to refer to our memory but instead we just process what we are looking at. The congruency effect shows that we do initially identify the scene's gist, and although it has a limited influence on the guidance of eye movements it does have an effect on the durations of fixations once a candidate object has been fixated, with fewer, longer fixations on incongruent objects that do not fit the gist of the scene.

References

1. Biederman, I.: Perceiving real-world scenes. Sci. 177, 77–80 (1972)
2. Biederman, I., Glass, A.L., Stacy, E.W.: On the information extracted from a glance at a scene. J. Exp. Psychol. 103, 597–600 (1973)
3. Biederman, I., Rabinowitz, J.C., Glass, A.L., Stacy, E.W.: On the information extracted from a glance at a scene. J. Exp. Psychol. 103, 597–600 (1974)
4. Davenport, J.L., Potter, M.C.: Scene consistency in object and background perception. Psychol. Sci. 15, 559–564 (2004)
5. Biederman, I., Mezzanotte, R.J., Rabinowitz, J.C.: Scene perception: Detecting and judging objects undergoing relational violations. Cog. Psychol. 14, 143–177 (1982)
6. Mackworth, N.H., Morandi, A.J.: The gaze selects informative details within pictures. Perc. Psychophys. 2, 547–552 (1967)
7. Loftus, G.R., Mackworth, N.H.: Cognitive determinants of fixation location during picture viewing. J. Exp. Psychol.: Hum. Perc. Perf. 4, 565–572 (1978)
8. De Graef, P., Christiaens, D., d'Ydewalle, G.: Perceptual effects of scene context on object identification. Psychol. Res. 52, 317–329 (1990)
9. Henderson, J.M., Weeks, P.A., Hollingworth, A.: The effects of semantic consistency on eye movements during complex scene viewing. J. Exp. Psychol.: Hum. Perc. Perf. 25, 210–228 (1999)
10. Itti, L., Koch, C.: A saliency-based search mechanism for overt and covert shifts of visual attention. Vis. Res. 40, 1489–1506 (2000)
11. Itti, L.: Quantitative modelling of perceptual salience at human eye position. Vis. Cog. 14, 959–984 (2006)
12. Parkhurst, D., Law, K., Niebur, E.: Modeling the role of salience in the allocation of overt visual attention. Vis. Res. 42, 107–123 (2002)
13. Underwood, G., Foulsham, T., van Loon, E., Underwood, J.: Visual attention, visual saliency, and eye movements during the inspection of natural scenes. In: Mira, J., Álvarez, J.R. (eds.) IWINAC 2005. LNCS, vol. 3562, pp. 459–468. Springer, Heidelberg (2005)
14. Underwood, G., Foulsham, T.: Visual saliency and semantic incongruency influence eye movements when inspecting pictures. Quart. J. Exp. Psychol. 59, 1931–1949 (2006)
15. Pomplun, M., Reingold, E.M., Shen, J.: Investigating the visual span in comparative search: the effects of task difficulty and divided attention. Cog. 81, 57–67 (2001)
16. Pomplun, M., Sichelschmidt, L., Wagner, K., Clermont, T., Rickheit, G., Ritter, H.: Comparative visual search: A difference that makes a difference. Cog. Sci. 25, 3–36 (2001)
17. Galpin, A.J., Underwood, G.: Eye movements during search and detection in comparative visual search. Perc. Psychophys. 67, 1313–1331 (2005)
18. Rensink, R.A.: To see or not to see: The need for attention to perceive changes in scenes. Psychol. Sci. 8, 368–373 (1997)
19. O'Regan, J.K., Rensink, R.A., Clark, J.J.: Change-blindness as a result of 'mudsplashes'. Nature 398, 334 (1999)
20. Stirk, J.A., Underwood, G.: Low-level visual saliency does not predict change detection in natural scenes. J. Vis. 7(10):3, 1–10 (2007)
21. Rensink, R.A.: The dynamic representation of scenes. Vis. Cog. 7, 17–42 (2000)
22. Hollingworth, A., Henderson, J.M.: Accurate visual memory for previously attended objects in natural scenes. J. Exp. Psychol.: Hum. Perc. Perf. 28, 113–136 (2002)
23. Hollingworth, A.: The relationship between online visual representation of a scene and long-term scene memory. J. Exp. Psychol.: Learn. Mem. Cog. 31, 396–411 (2005)
24. O'Regan, J.K.: Solving the "real" mysteries of visual perception: The world as an outside memory. Canad. J. Psychol. 46, 461–488 (1992)

Spatiotemporal Saliency: Towards a Hierarchical Representation of Visual Saliency

Neil D.B. Bruce and John K. Tsotsos

Department of Computer Science and Engineering and
Centre for Vision Research
York University, Toronto, ON, Canada
{neil,tsotsos}@cse.yorku.ca
http://www.cse.yorku.ca/~neil

Abstract. In prior work, we put forth a model of visual saliency motivated by information theoretic considerations [1]. In this effort we consider how this proposal extends to explain saliency in the spatiotemporal domain and further, propose a distributed representation for visual saliency comprised of localized hierarchical saliency computation. Evidence for the efficacy of the proposal in capturing aspects of human behavior is achieved via comparison with eye tracking data and a discussion of the role of neural coding in the determination of saliency suggests avenues for future research.

Keywords: Attention, Saliency, Spatiotemporal, Information Theory, Fixation, Hierarchical.

1 Introduction

Certain visual search experiments demonstrate in dramatic fashion the immediate and automatic deployment of attention to unique stimulus elements in a display. This phenomenon no doubt factors appreciably into visual sampling in general influencing fixational eye movements and our visual experience as a whole. Some success has been had in emulating these mechanisms [2], reproducing certain behavioral observations related to visual search, but the precise nature of the principles underlying such behaviors remains unknown.

One recent proposal deemed Attention by Information Maximization (AIM) is grounded in a principled definition for what constitutes visually salient content derived from information theory, and has had some success in explaining certain aspects of behavior including the deployment of eye movements [1] and other visual search behaviors [3]. In this paper we further explore support for this proposal through consideration of spatiotemporal visual stimuli. This includes a comparison of the proposal against the state of the art in this domain. The following discussion reveals the efficacy of the proposal put forth in AIM to explain eye movements for spatiotemporal data and also describes how the model

L. Paletta and J.K. Tsotsos (Eds.): WAPCV 2008, LNAI 5395, pp. 98–111, 2009.

fits in with the *big picture*. Specifically, we address how the proposal fits with distributed hierarchical attentional architectures of the sort put forth by Tsotsos [4] for which favorable evidence has appeared in recent years.

2   AIM: Information Maximizing Saliency

In the following section, we briefly review the proposal put forth in [1], which is applied to a set of neurons that code for content in space-time within the evaluation included in this work. The following offers only a brief overview; for a detailed account, readers should refer to [1].

The central premise of AIM is that saliency computation should serve to maximize information sampled from one's environment from a stimulus driven perspective. Specifically, given an ensemble of neurons $C_{i,j}$ that code for content at spatial coordinates i, j with $C_{i,j,k}$, $k = 1...N$ corresponding to the different types of cells with receptive fields centered at i, j the self-information or surprisal associated with $C_{i,j}$ is given by $-log(p(C_{i,j}))$ with the likelihood determined by observing the response of cells in the surround of $C_{i,j}$. Given the assumption of independence on the response of different types of cells (an assumption made reasonable by sparsity as discussed in the section that follows), this quantity may be computed as $\sum_{k=1}^{N} -log(C_{i,j,k})$. Saliency in this context then amounts to the surprisal or self-information of the response associated with a cell as defined by its surround. In other words, saliency is inversely proportional to the likelihood of predicting the response of any given neuron in observing the response of neurons in its surrounding spatiotemporal context. For any given cell type it is straightforward to derive a likelihood estimate by constructing a probability density estimate based on cells of the same type in the surround. An overview of the model with reference to the specifics of the implementation for spatiotemporal stimuli is presented in the section that follows.

3   Extension to Space-Time

The general nature of the original proposal implies that it may be applied to any set of neurons that constitute a sparse basis. For this reason, extension to space-time is straightforward assuming the early coding of spatiotemporal content observed in the cortex satisfies these criteria. There exist many efforts documenting the relationship between early visual cortical neurons and coding strategies that demonstrate that learning a sparse code for local grey-level image content yields V1 like receptive fields similar to oriented Gabor filters [5,6]. Further efforts have demonstrated this same strategy yields color-opponent coding for spatiochromatic content [7] and also cells with properties akin to V1 for spatiotemporal data [8]. We have employed the same data and strategy put forth in [8] to learn a basis set of cells coding for spatiotemporal content. The data described in [8] was subsampled taking every second frame to yield data at 25 frames per second. The data set consists of a variety of natural spatiotemporal sequences taken from various angles of a moving vehicle traveling in a typical urban environment. Spatiotemporal volumes were then randomly sampled from the videos to yield 11x11x6 (x,y,t) localized spatiotemporal volumes that served as training data. Infomax ICA [9] was applied to the training set resulting in a spatiotemporal basis consisting of cells that respond to various frequencies and velocities of motion and for which the correlation between cell firing rates is minimized. The basis resulting from dimensionality reduction via PCA retaining 95% variance followed by ICA yields a set of 60 spatiotemporal cells. A subsample of these (corresponding to 1st, 3rd and 6th frame of the volume) are shown in figure 1. Note the response to various angular and radial frequencies and selectivity for different velocities of motion. Aside from the application to spatiotemporal data and the different basis set, the saliency computation proceeds according to the description put forth in [1].

An overall schematic of the model based on the learned spatiotemporal basis appears in figure 2. A localized region from adjacent frames (3 of 6 shown) are projected onto the learned basis. This yields a set of coefficients for the local region that describes the extent to which various types of motion are observed at the given location. The likelihood of each response is then evaluated by observing the response of cells of the same type in the surround or in this implementation, over the entire image. A sum of the negative log likelihood associated with all of the coefficients corresponding to the given coordinate (pixel) location yields a local measure of saliency.

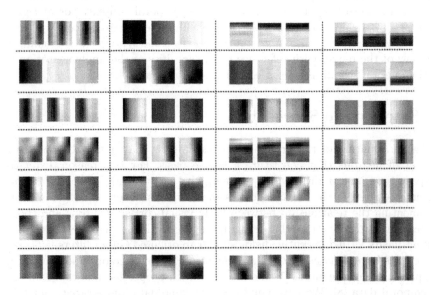

Fig. 1. The receptive field profile of a subsample of the learned basis corresponding to frames 1, 3 and 6 of the spatiotemporal volume. Note the selectivity for various angular and radial frequencies and velocities and directions of motion.

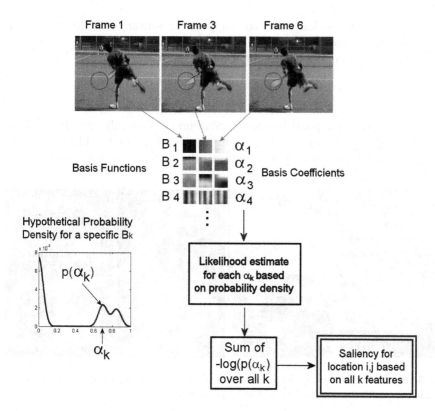

Fig. 2. An overview of the computation performed by AIM. A spatiotemporal volume is projected onto a learned basis based on independent component analysis. The likelihood of any given cells firing rate may be estimated by observing the distribution of responses associated with cells of the same type in the surround or over the entire image. A summation of these likelihoods subjected to a log transform then yields a local measure of information. For a complete description the reader should refer to [1].

4   Evaluation

An evaluation of the efficacy of the model in predicting spatiotemporal fixation patterns is achieved via comparison with eye tracking data collected for video stimuli. The eye tracking data employed for this study was that used in [10] and performance evaluation was carried out according to the same performance metric described in the aforementioned work.

The data consists of eye tracking data for a total of 50 video clips and from 8 subjects aged 22-32 with normal or corrected to normal vision. Videos consist of indoor and outdoor scenes, news and television clips and video games. Videos were presented at a resolution of 640x480 and at 60 Hz and consist of over 25 minutes of playtime. The total number of saccades included in the analysis is 12,211.

For any given algorithm, one may compare the saliency at fixated locations with randomly sampled locations. The Kullback-Leibler divergence of two distributions corresponding to these quantities is given by

$$D_{KL}(P,Q) = \sum P(i)log\frac{P(i)}{Q(i)}$$

where P and Q correspond to the distribution of randomly sampled and at-fixation sampled saliency values respectively based on 10 bin histogram estimates. The KL-divergence offers a performance metric allowing comparison of

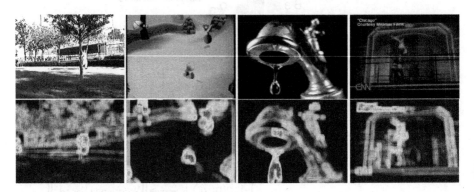

Fig. 3. Relative saliency of each pixel for a variety of frames from different videos allowing a qualitative assessment of model performance

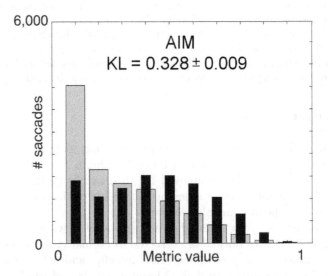

Fig. 4. A histogram representation comparing saliency values at fixated versus randomly located display locations. KL-divergence is 0.328 as compared with 0.241 for the algorithm presented in [10] and 0.205 for that appearing in [2].

various algorithms. Results are compared against those put forth in [10] and proceeds according to the same performance evaluation strategy.

Figure 3 demonstrates the relative saliency of pixel locations for a variety of single frames from a number of videos. Note the inherent tradeoff between moving and stationary content as observed for the running tap, and park scene as well as the ability to detect salient patterns on a relatively low contrast background (rightmost frame).

Figure 4 demonstrates a histogram of the saliency associated with the fixated locations as compared with those from uniformly randomly sampled regions. Of note is the shift of the distribution towards higher saliency values for the distribution associated with fixated relative to random locations. The KL-divergence of the two distributions shown is 0.328. This compares favorably with the Surprise metric of Itti and Baldi [10] which gives rise to a KL-divergence score of 0.241 and the saliency evaluation of Itti and Koch [2] which yields a KL-divergence score of 0.205. This result demonstrates that relative to competing proposals the saliency associated with fixated relative to random locations is greatest for AIM.

5 Surround Suppression, Gain Control and Redundancy

An important consideration in any model that posits a specific proposal for how saliency computation is achieved, is that of a possible neural implementation. Perhaps the foremost consideration pertaining to neural circuitry, is the extent to which the proposal agrees with observations concerning cortical circuitry and neurophysiology. To this end, this section reviews a variety of classic and recent results derived from psychophysics and imaging experiments on the nature of surround suppression within the cortex. Necessary conditions of an architecture that seeks to maximize information in its control of neural gain are weighed against the experimental literature in order to evaluate the plausibility of AIM from the perspective of a possible neural basis for its implementation. As a whole, the discussion establishes that a variety of peculiar and very specific constraints imposed by the implementation show considerable agreement with the computation implicated in surround suppression further providing support for AIM, and also offering some insight on the nature of computation responsible for iso-orientation surround suppression in early visual cortex. Debate concerning the specific nature and form of surround suppression has rekindled in recent years, which has resulted in a large body of interesting results that further elucidate the details of this process. The following discussion reviews these results and offers further insight through a meta-analysis of recent studies. In each case, experimental findings are contrasted against the computational constraints on AIM to establish plausibility of the proposed computation.

5.1 Types of Features

A great deal of research has focused specifically on the suppression that arises from introducing a stimulus in the surround of a localized oriented Gabor target. The specific nature of iso-orientation (iso-feature) surround suppression as dictated by the details of AIM includes two key considerations: 1. Suppression of a cell whose receptive field lies at the target location should occur only for a surround stimulus that is the effective stimulus for this cell. For example, for a vertically oriented Gabor target, suppression of a cell that elicits a response to the target will occur only by way of a similar stimulus appearing in the surround. Recall that a fundamental assumption is that the responses of different types of cells at a given location are such that the correlation between their responses is minimal and this is a phenomenon that is observed cortically. In the domain of studies pertaining to surround suppression, the literature is undivided in its agreement with this assumption. When considering the cell response or psychometric threshold associated with a target patch, suppression from a surround stimulus is highly stimulus specific and is at a maximum for a surround matching the target orientation, with suppression observed only for a narrow orientation band centered around the target orientation [11,12,13,14,15,16]. This is consistent with a local likelihood estimate in which the independence assumption is implicit. 2. Suppression should be observed for all feature types, and the nature of, and parameters associated with suppression should not differ across feature type. This is an important consideration since studies of this type have largely focused on oriented sinusoidal stimuli but nevertheless similar suppression associated with color, or velocity of motion for example, should also be observed and the nature of such suppression should be consistent with that observed in studies involving oriented sinusoidal target and surrounds. One recent effort provides strong evidence that this is the case through single cell recording on macaque monkeys [14]. Shen et al. demonstrate that centre-surround fields defined by a variety of features including color, velocity and oriented gratings all elicit suppression and with suppression at a maximum for matching centre and surround stimuli.

5.2   Relative Contrast

Given a cell with firing rate $N_{i,j}$ that codes for a specific quantity at coordinates i,j in the visual field (e.g. a cell selective for a specific angular and radial frequency as part of a basis representation with its centre at location i,j), a density estimate on the observation likelihood of the firing rate associated with $N_{i,j}$ as discussed earlier in this section is given by:

$$p(N_{i,j}) = \sum_{\forall s,t \in \Omega} f(N_{i,j} - N_{s,t}) \tag{1}$$

Where f is a monotonic symmetric kernel with its maximum at f(0) and Ω the region over which the surround has any significant impact. For further ease of exposition in observing the behavior of equation 1, assume without loss of generality that f comprises a Gaussian kernel. Then equation 1 becomes:

$$\frac{1}{\sigma\sqrt{2\pi}} \sum_{\forall s,t \in \Omega} e^{-(N_{j,k}-N_{s,t})^2/2\sigma^2} \tag{2}$$

As there also exists a spatial component to this estimate, it may be more appropriate to also include a parameter that reflects the effect of distance on the contribution of any given cell to the estimate of $N_{i,j}$ which might appear as follows:

$$\frac{1}{\sigma\sqrt{2\pi}} \sum_{\forall s,t \in \Omega} \Psi(s,t)e^{-(N_{j,k}-N_{s,t})^2/2\sigma^2} \tag{3}$$

Ψ drops off according to the distance of any given cell from the target location, reflecting the decreasing correlation between responses. Assuming that surround suppression is the basis for the computation involved in AIM equation 1 demands a very specific form for the suppressive influence of a surrounding stimulus on the target item. According to the form of equation 3, suppression depends on the relative response of centre and surround stimuli and should be at a maximum for equal contrast centre and surround stimuli: Raising or lowering the contrast of a stimulus pattern will generally result in a concomitant increase in the response of a cell for which the pattern in question is the effective stimulus. There is therefore a direct monotonic (nonlinear) relationship between the firing rate attributed to centre or surround, and their respective contrasts. Support for suppression as a function of relative centre versus surround contrast is ubiquitous in the literature [17,18,14,11,19,20,15,21] although there is as of yet no consensus on why this should be the specific form for the suppressive influence of a surround stimulus. There also exists a large body of prominent studies revealing that this suppression is indeed at a maximum for equal contrast centre and surround stimuli [17,18,14,11,15]. Note that this implies mathematical equivalence between surround suppression and a likelihood estimate on a given cell's response as defined by the response of neighboring cells and implies divisive modulation of a cells response by a function of its likelihood. This is an important consideration as it offers insight on the role of surround suppression which has recently become an issue of considerable dispute [16] and implicates surround suppression as the machinery underlying the implementation of AIM. It is also worth noting that the suppressive impact of cells in the surround is observed to drop off exponentially with distance from the target giving the specific form of Ψ [16].

5.3   Spatial Configuration

For the sake of exposition, let us assume that the computation under discussion is restricted to V1. From the perspective of efficient coding, no knowledge of structure is available at V1 beyond that which lies within a region the size of single V1 receptive field. A pure information theoretic interpretation of the surprisal associated with a local observation as determined at the level of V1 should reflect this implying an isotropic contribution to any likelihood estimate in the vicinity of the target cell, regardless of the pattern that forms an effective stimulus for the cell in question. That is, for a unit whose effective stimulus is a horizontal Gabor pattern, equidistant patterns of the same type in the vicinity of the target should result in equal suppression regardless of where they appear with respect to the target and this is reflected in the implementation put forth in [1]. It is also expected that likelihoods associated with higher order structure over larger receptive fields are mediated by higher visual areas either implicitly at the single cell level or explicitly via recurrent connections. In line with the assumption that computation is on the observation likelihood of a pattern within a given region, and that structures are limited to an aperture no larger than a V1 receptive field, it is indeed the case that suppression from the surround is isotropic with respect to the location of a pattern appearing in the surround independent of target and surround orientations [16]. By virtue of the same consideration, one would also expect the spatial extent of surround suppression to be invariant to the spatial frequency of a target item. This is also a consideration that is evident in the literature [16]. In consideration of observation likelihoods associated with more complex patterns, it is interesting to consider the nature of surround suppression among higher visual areas. Recent studies are discovering more and more examples of suppressive surround inhibition among higher visual areas with the same properties and divisive influence as those that are well established in V1. Extrastriate surround inhibition of this form has been observed at least among areas V2 [22,23], V4 [24,25], MT [26,27,28], and MST [29]. This is suggestive of the possibility that saliency is represented within a distributed hierarchy, with local saliency computation mediated by surround suppression at various layers of the visual cortex.

5.4   Fovea versus Periphery

If the role of local surround suppression is in attenuating neural activation associated with unimportant visual input and/or redirecting the eyes via fixational eye movements one would expect the influence of such a mechanism to be prominent within the periphery of the visual field. Petrov and McKee demonstrated that surround suppression is in fact strong in the periphery and absent in the fovea [16]. This is consistent, as Petrov and McKee point out, with a role of this mechanism in the control of saccadic eye movements. Furthermore, there are additional points they highlight that support this possibility, including the fact that the extent of suppression is invariant to stimulus spatial frequency. Also of note, is the fact that the inaccuracy of a first saccade is proportional to target eccentricity and this correlates with the extent of surround suppression as a function of eccentricity [16]. Note that the cortical region over which surround suppression is observed does not vary with eccentricity implying that computationally, an equal number of neurons contribute to any given likelihood estimate of the form appearing in equation 1. All of these considerations are in line with a role of this mechanism in the deployment of saccades.

5.5   Summary

We have put forth the proposal that the implementation of AIM is achieved via local surround circuitry throughout the visual cortex. As a whole, there appears to be considerable agreement with the proposal and the specific form of surround suppression. The demonstration of equivalence of a likelihood estimate on the surround of a cell with the apparent form of suppressive inhibition implies modulation of cell responses at a single cell level through divisive gain as a function of the likelihood associated with that cell's response. This provides a more specific explanation for the nature of computation appearing in suppressive surround circuitry and further bolsters the claim that saliency computation proceeds according to a strategy of optimizing information transmission.

6   On the Role of Neural Encoding

As discussed, probability density estimation, or any sort of neural probabilistic inference, requires an efficient representation of the statistics of the natural world in order to meet computational demands. The specific nature of this representation within many biological brains seems to be an encoding of natural stimuli in a manner that minimizes the correlation or mutual dependence between neurons [30,31,32,33,34]. A consequence of this computationally is that likelihoods in regard to a neural firing rate can be considered independent of the firing rates of neurons that code for different features. In this regard, the pop-out versus serial search distinction may be seen as an emergent property of this coding strategy. Since likelihoods associated with orientation statistics are considered independently of those that represent chromatic information, the conjunction of these features fails to elicit pop-out [3]. It is also interesting to note in support of this line of reasoning, that as radial and angular frequency are coded jointly within the cortex, a unique item defined by a conjunction of spatial frequency and orientation does result in a pop-out stimulus [35]. In light of this observation, it may be said more generally, that the specific nature of neuron properties has a considerable influence on the behavior that manifests. It is well established that search efficiency is more involved than a simple dichotomy of serial versus parallel searches [36]. It has been demonstrated that one can observe a wide range of behaviors from very efficient to very inefficient depending on the chosen stimuli. One might suggest that the extent to which a search may be carried out efficiently reflects the complexity of the neural code corresponding to target and distractor elements. For stimuli that are highly natural and may be represented by the response of a small number of neurons, one might expect a far more efficient search than that associated with a highly unnatural stimulus that gives rise to a widely distributed neural representation. This may also extend beyond simple V1-like features to explain the surprising efficiency with which some search tasks involving complex stimuli are completed, such as search tasks involving 3D-shape [37], depth from shading [38] and even very complex forms such as faces [39] which are known to have a highly efficient cortical representation within the primate cortex [40,41,42]. Considerations pertaining to coding may also shed some light on the role of novelty in determining search efficiency. Inter-element suppression of stimulus items may occur more strongly for those representations that are relatively efficient and carried by only a small number of cells. Behaviorally this is consistent with visual search paradigms in which familiarity with distractors yields a relatively efficient search [43,44] assuming familiarity with target items leads to a more efficient or even template like representation of the relevant stimuli. As a whole, it may be said that the role that principles underlying coding within the visual cortex play within attention and visual search is an aspect of the problem that has been underemphasized. Many behaviors, in particular the specific efficiency with which a search is conducted, may be seen as properties that surface from very basic principles underlying the neural representation of visual patterns, and consideration of the specific role of coding in attention and visual search should serve as a target for further investigation.

7   Towards a Hierarchical Representation of Saliency

The preceding results demonstrate that the proposal originally tested on spatiochromatic data extends well to explain spatiotemporal data. A question that naturally follows from this, is the extent to which the proposal may extend to capture more high-level behaviors associated with neurons coding for more complex stimuli and appearing higher in the cortex. As the saliency associated with a pixel location is a simple summation of the individual saliency attributed to each cell for each location, it is evident that saliency may be evaluated at the level of a single cell. It follows that the same proposal that has been depicted in a form more akin to the traditional saliency map style representation may also reside within a distributed hierarchical representation in which the representation of saliency is implicit and computed via local modulation as opposed to a single explicit topographical representation of saliency. Such a proposal is in line with models of attention that posit a distributed hierarchical selection strategy [4]. Additionally, as the constraints on the cells involved are satisfied among higher visual areas, one might propose that the proposal put forth in AIM extends to higher visual areas to explain some of the apparent *high-level* effects documented in the previous section. For example, a hierarchical coding structure combined with AIM should afford some of the pop-out effects associated with high-level features such as depth from shading assuming an appropriate code for such features among higher visual areas.

8   Conclusion

We have considered how AIM extends to capture behaviors associated with visual patterns distributed over space and time. The plausibility of the proposal as a description of human behavior is validated through a comparison with eye tracking data on a wide range of qualitatively different videos. The proposal emerges as very effective in explaining the behavioral data as was demonstrated for the spatiochromatic case. We have also described how the proposal put forth in AIM is compatible with distributed architectures for attentional selection [4] including related details pertaining to coding and neural implementation. This is an important contribution as the topic of saliency [4] is seldom discussed in a context independent of the assumption of an explicit topographical saliency map. Future work will aim to further explore saliency computation as a process involving attention acting on a distributed hierarchical representation with saliency realized via localized modulation throughout the cortex.

Acknowledgments. The authors wish to thank Dr. Laurent Itti for sharing the eye tracking data employed in the evaluation of spatiotemporal saliency. The authors gratefully acknowledge the support of NSERC in funding this work. John Tsotsos is the NSERC Canada Research Chair in Computational Vision.

References

1. Bruce, N.D.B., Tsotsos, J.K.: Saliency Based on Information Maximization. In: Advances in Neural Information Processing Systems, vol. 18, pp. 155–162 (June 2006)
2. Itti, L., Koch, C., Niebur, E.: A Model of Saliency-Based Visual Attention for Rapid Scene Analysis. IEEE Transactions on Pattern Analysis and Machine Intelligence 20(11), 1254–1259 (1998)
3. Bruce, N.D.B., Tsotsos, J.K.: An information theoretic model of saliency and visual search. In: Paletta, L., Rome, E. (eds.) WAPCV 2007. LNCS, vol. 4840, pp. 171–183. Springer, Heidelberg (2007)
4. Tsotsos, J.K., Culhane, S., Yan Kei Wai, W., Lai, Y., Davis, N., Nuflo, F.: Modeling visual attention via selective tuning. Artificial intelligence 78, 507–545 (1995)
5. Bell, A.J., Sejnowski, T.J.: The 'Independent Components' of Natural Scenes are Edge Filters. Vision Research 37(23), 3327–3338 (1997)
6. Olshausen, B.A., Field, D.J.: Emergence of simple-cell receptive field properties by learning a sparse code for natural images. Nature 381, 607–609 (1996)
7. Wachtler, T., Lee, T.-W., Sejnowski, T.J.: The chromatic structure of natural scenes. J. Opt. Soc. Amer. A 18(1), 65–77 (2001)
8. van Hateren, J.H., van der Schaaf, A.: Independent component filters of natural images compared with simple cells in primary visual cortex. Proc. R. Soc. Lond. B 265, 359–366 (1998)
9. Lee, T.W., Girolami, M., Sejnowski, T.J.: Independent component analysis using an extended infomax algorithm for mixed subgaussian and supergaussian sources. Neural Computation 11(2), 417–441 (1999)
10. Itti, L., Baldi, P.: Bayesian Surprise Attracts Human Attention. In: Advances in Neural Information Processing Systems, vol. 19, pp. 547–554 (2006)
11. Yu, C., Levi, D.M.: Surround modulation in human vision unmasked by masking experiments. Nature 3(7), 724–728 (2000)
12. Williams, A.L., Singh, K.D., Smith, A.T.: Surround modulation measured with fMRI in the visual cortex. Journal of Neurophysiology 89(1), 525–533 (2003)
13. Xing, J., Heeger, D.J.: Measurement and Modeling of Centre-Surround Suppression and Enhancement. Vision Research 41, 571–583 (2001)
14. Shen, Z.M., Xu, W.F., Li, C.Y.: Cue-invariant detection of centre surround discontinuity by V1 neurons in awake macaque monkey. Journal of Physiology 583, 581–592 (2007)
15. Yu, C., Klein, A.K., Levi, D.M.: Cross-and Iso-oriented surrounds modulate the contrast response function: The effect of surround contrast. Journal of Vision 3, 527–540 (2003)
16. Petrov, Y., McKee, S.P.: The effect of spatial configuration on surround suppression of contrast sensitivity. Journal of Vision 6(3), 224–238 (2006)
17. Adini, Y., Sagi, D.: Recurrent networks in human visual cortex: psychophysical evidence. Journal of the Optical Society of America A 18(8), 2228–2236 (2001)

18. Olzak, L.A., Laurinen, P.I.: Contextual Effects in fine spatial discriminations. Nature 381(6583), 607–609 (2005)
19. Cannon, M.W., Fullencamp, S.C.: A model for inhibitory lateral interaction effects in perceived contrast. Vision Research 36(8), 1115–1125 (1996)
20. Xing, J., Heeger, D.J.: Centre-surround interactions in foveal and peripheral vision. Vision Research 40, 3065–3072 (2000)
21. Yu, C., Klein, A.K., Levi, D.M.: Surround modulation of perceived contrast and the role of brightness induction. Journal of Vision 1, 18–31 (2001)
22. Zhang, B., Zheng, J., Watanabe, I., Maruko, I., Bi, H., Smith, E.L., Chino, Y.: Delayed maturation of receptive field centre/surround mechanisms in V2. Proceedings of the National Academy of Sciences 102(16), 5862–5867 (2005)
23. Solomon, S.G., Pierce, J.W., Lennie, P.: The impact of suppressive surrounds on chromatic properties of cortical neurons. Journal of Neuroscience 24(1), 148–160 (2004)
24. Schein, S.J., Desimone, R.: Spectral properties of V4 Neurons in the macaque. Journal of Neuroscience 10(10), 3369–3389 (1990)
25. Kondo, H., Komatsu, H.: Suppression on neuronal responses by a metacontrast masking stimulus. Neuroscience Research 36(1), 27–33 (2000)
26. Tadin, D., Lappin, J.S.: Optimal Size for perceiving motion decreases with contrast. Vision Research 45, 2059–2064 (2005)
27. Born, R.T., Bradley, D.C.: Structure and Function of Visual Area MT. Annual Review of Neuroscience 28, 157–189 (2005)
28. Huang, X., Albright, T.D., Stoner, G.R.: Adaptive Surround Modulation in Cortical Area MT. Neuron 53(5), 761–770 (2007)
29. Eifuku, S., Wurtz, R.H.: Response to Motion in Extrastriate Area MSTl: Centre-Surround Interactions. Journal of Neurophysiology 80(11), 282–296 (1998)
30. Foldiak, P., Young, M.: Sparse coding in the primate cortex. In: Arbib, M.A. (ed.) The Handbook of Brain Theory and Neural Networks, pp. 895–898 (1995)
31. David, S.V., Vinje, W.E., Gallant, J.L.: Natural stimulus statistics alter the receptive field structure of v1 neurons. Journal of Neuroscience 24(31), 6991–7006 (2004)
32. Simoncelli, E.P., Olshausen, B.A.: Natural image statistics and neural representation. Annual Review Neuroscience 24, 1193–1216 (2001)
33. Quian Quiroga, R., Reddy, L., Kreiman, G., Koch, C., Fried, I.: Invariant visual representation by single neurons in the human brain. Proceedings of the National Academy of Science 102(16), 5862–5867 (2005)
34. Kreiman, G.: Neural coding: computational and biophysical perspectives. Physics of Life Reviews 2, 71–102 (2004)
35. Sagi, D.: The combination of spatial frequency and orientation is effortlessly perceived. Perception and Psychophysics 43, 601–603 (1988)
36. Wolfe, J.M., Horowitz, T.S.: What attributes guide the deployment of visual attention and how do they do it? Nature Reviews Neuroscience 5, 1–7 (2004)
37. Enns, J.T., Rensink, R.A.: Sensitivity to three-dimensional orientation in visual search. Psychological Science 1, 323–326 (1990)
38. Ramachandran, V.S.: Perception of Shape from Shading. Nature, 163–166 (1988)
39. Hershler, O., Hochstein, S.: At first sight: a high-level pop out effect for faces. Vision Research 45(13), 1707–1724 (2005)
40. Sergent, J., Ohta, S., MacDonald, B.: Functional neuroanatomy of face and object processing. A positron emission tomography study. Brain 115(1), 15–36 (1992)

41. Kanwisher, N., McDermott, J., Chun, M.M.: The fusiform face area: a module in human extrastriate cortex specialized for face perception. Journal of Neuroscience 17(11), 4302–4311 (2006)
42. Grill-Spector, K., Sayres, R., Ress, D.: High-resolution imaging reveals highly selective nonface clusters in the fusiform face area. Nature Neuroscience 9(9), 1177–1185 (2006)
43. Wang, Q., Cavanagh, P., Green, M.: Familiarity and pop-out in visual search. Perception and Psychophysics 56(5), 495–500 (1994)
44. Shen, J., Reingold, E.M.: Visual search asymmetry: the influence of stimulus familiarity and low-level features. Perception and Psychophysics 63(3), 464–475 (2001)

Motion Saliency Maps from Spatiotemporal Filtering

Anna Belardinelli, Fiora Pirri, and Andrea Carbone

Dipartimento di Informatica e Sistemistica
Sapienza University of Rome
{belardinelli,pirri,carbone}@dis.uniroma1.it

Abstract. For artificial systems acting and perceiving in a dynamic world a core ability is to focus on aspects of the environment that can be crucial for the task at hand. Perception in autonomous systems needs to be filtered by a biologically inspired selective ability, therefore attention in dynamic settings is becoming a key research issue.

In this paper we present a model for motion salience map computation based on spatiotemporal filtering. We extract a measure of coherent motion energy and select by the center-surround mechanism relevant zones that accumulate most energy and therefore contrast with surroundings in a given time slot.

The method was tested on synthetic and real video sequences, supporting biological plausibility.

1   Introduction

Visual attention has emerged in recent years as a powerful tool to make computer and robot vision more and more effective in a great variety of tasks, since it allows to focus analysis and processing on some restrained portions of images and frames. With most approaches being biologically inspired [1],[2],[3], attention computational modelling has made it possible to introduce a basic cognitive level between sensor data collection and perception interpretation for high level reasoning. The design of an artificial system required to navigate, act or reason in a dynamic world and to interact with other partners, interpreting their intentions and movements, has to take into account in the attention architecture a mechanism for identifying and selecting significant motion.

Sensitivity to visual motion is of course extremely important in terms of survival for any evolved biological system, be it a prey or a predator. This is deeply rooted in the neurology of the brain since cells in different areas of the primate cortex are assigned to detection of different motion patterns and velocities (as explained and modelled in [4]). Other models of motion sensing and perception according to human psychophysics have made use of spatiotemporal filtering. Adelson and Bergen [5] showed how such models enable motion detection and direction selection in terms of spatiotemporal oriented energy, for both continuous and sampled motion. Heeger [6], starting from similar premises, extracted velocity information to obtain optical flow and compared it with physiological processes in humans. Watson and Ahumada [7] modelled a motion sensor, meant to provide quantitative information about physical stimuli displaying motion. Simoncelli [8] presented a Bayesian framework for motion estimation based on brightness conservation.

L. Paletta and J.K. Tsotsos (Eds.): WAPCV 2008, LNAI 5395, pp. 112–123, 2009.

As to attention modulation, [9] visual motion is said to naturally draw attention in a pure bottom-up fashion only if it is related to a new object onset that motion helps segregating. During visual search, motion is effortlessly detected if, like other pre-attentive features, it denotes a target, thus causing a pop out effect. That is, contrast to local dynamics in the scene attracts attention, like color or intensity contrast do. This explains why search for a moving target among stationary distractors is easier than in the opposite case or than searching for a faster moving target among slow moving distractors. Basically, as shown in [10], the more the velocity of an element differs from that of the surrounding the more the element is salient. In that work motion salience and thus pop-out was defined by the amount of Mahalanobis distance between target velocity and the mean velocity of the distractors.

Attention models or vision architectures and applications have often taken into account motion as an informative feature to detect and segment interesting objects or targets by means of optical flow computation, block matching or other motion detection techniques [11],[12],[13],[14]. These methods consider principally differences between subsequent frames, not accounting for a broader analysis of motion.

In this paper we apply spatiotemporal filtering to attention focus selection in dynamic scenes. This method allows for motion evolution in the space-time domain, thus highlighting motion patterns as particular energy signatures in 2D planes slicing the space-time volume [15]. The link from this smart motion detection to attention is achieved by means of center-surround and maxima selection enhancing areas of salient motion.

In next sections we first review spatiotemporal analysis of sequences, show then how we used it to obtain selective saliency maps and applied it to synthetic and natural video sequences, showing selection of locations characterized by high levels of energy motion along a given time unit.

2 Spatiotemporal Energy for Coherent Motion Distinction

As mentioned in Section 1 analysis of motion through spatiotemporal filtering has been widely investigated by works on the perception of motion [5], [7], [6] and further developed more recently by [15]. These models showed how motion analysis based on low level visual information is particularly straightforward in the spatiotemporal domain. Specifically analysis is performed on $(x - y - t)$ volumes, obtained by collecting together a set of consecutive frames. This representation allows analysis on a longer time horizon rather than from frame to frame, leading to discrimination and qualitative understanding of the unfolding of perceived motion. Moreover it has the advantage of avoiding matching or correspondence search problem.

An $x - y - t$ volume can be considered as constituted by a stack of 2D $x - t$ planes or by aligning vertically 2D $y - t$ planes. In these planes motion is represented by slanted tracks whose slope is proportional to velocity. Thus detecting motion in the spatiotemporal representation reduces to detecting oriented edges in 2D planes corresponding to rows and columns of the frames composing the volume [5]. This can be done in low level processing by following some basic steps we explain below.

In [16] and [15] oriented spatiotemporal energy is used to obtain a measure of motion salience with respect to some important directions characterizing coherent motion.

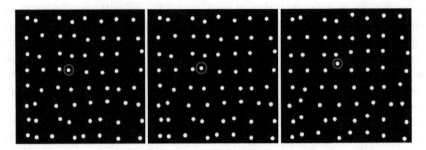

Fig. 1. An example of visual popout: frames at time $t, t + 5, t + 10$. The faster dot is highlighted by a red circle. The considered sequence was $256 \times 256 \times 40$.

This latter is indeed defined by high energy derived by response to filters oriented along spatiotemporal diagonal directions, while other insignificant types of motion (static, unstructured, incoherent, flicker, scintillation) result in low energy along these directions. According to indications on neurophysiological plausibility in [5] and [6], we use Gabor filters to extract motion information concerning horizontal and vertical motion within a specific spatiotemporal frequency, i.e. (in $x - t$):

$$G^o_{\theta_i}(x, t) = \frac{1}{2\pi\sigma_x\sigma_t} exp[-(\frac{x^2}{2\sigma_x^2} + \frac{t^2}{2\sigma_t^2})] \sin(2\pi\omega_{x_0} x + 2\pi\omega_{t_0} t) \qquad (1)$$

where ω_v, σ_v, with $v \in \{x, t\}$, include the filter parameters encompassing the receptive field linear dimension and the wavelength, while θ_i denotes the selected orientation obtained suitably rotating the filter. Analogous formulation applies to the other spatiotemporal dimension (y, t).

We refer to a particular spatiotemporal scale but the proceeding is extendible to multiple spatial directions and spatial and time frequencies to capture a wider spectrum of motion information and compute salience on more channels.

For example, consider a video sequence [1] where a motion pop-out is displayed. Here 64 white dots describe a circular orbit against a black background (see some snapshots in Fig.1). Each dot has different phase but same period, except for the dot in position $(4, 4)$ (referring to rows and columns of moving dots) that stands out because of its larger velocity.

Coherent energy computation for a given space-time volume goes like this :

1. **Gabor filtering and oriented energy computation.** Each plane $I(x, t)$ and $I(y, t)$ is filtered by the 2D Gabor filter (1) at $\theta_1 = 45^{\text{deg}}$ and $\theta_2 = 135^{\text{deg}}$, extracting leftward/rightward and downward/upward motion respectively. Filters are taken in quadrature (odd and even) and responses to quadrature pairs are squared and summed to obtain independence from phase. We hence obtain the following energies for righ/leftward (ER,EL) motion for any fixed y_i:

$$EL_{y_i}(x, t) = (G^o_{\theta_1}(x, t) * I(x, t))^2 + (G^e_{\theta_1}(x, t) * I(x, t))^2 \qquad (2)$$

$$ER_{y_i}(x, t) = (G^o_{\theta_2}(x, t) * I(x, t))^2 + (G^e_{\theta_2}(x, t) * I(x, t))^2 \qquad (3)$$

---

[1] http://www.scholarpedia.org/article/Image:VisualSalience_Motion.gif

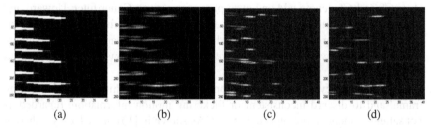

(a)   (b)   (c)   (d)

Fig. 2. (a): The $y - t$ plane related to column 22, i.e. evolution along time of column 22 for the sequence of Fig.1, with $y = 1 : 256$ (ordinates) and $t = 1 : 40 frames \sim 2sec.$ (abscissas). (b): Extraction of downward edges via Gabor filtering. (c), (d): representation of upward and downward energies for the considered plane.

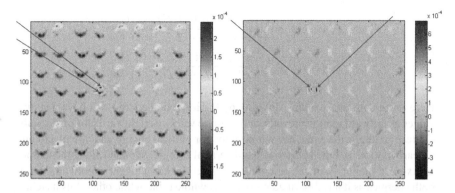

Fig. 3. Left: horizontal energy summed 2 up to sec. (about 40 frames). Right: Vertical energy summed up to 2 sec. (about 40 frames). Both figures refer to the example of Fig. 1 and 2.

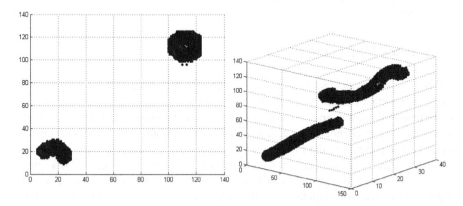

Fig. 4. Comparison of energy patterns between the faster dot of the example illustrated in Fig.1 and one of the other dots. Left:frontal view (x,y axes). Right: rotated view (x,y,t axes, t=1:40 frames).

Here G^o and G^e denote odd and even phase for a given Gabor filter. Upward and downward edge extraction is achieved analogously by filtering planes $I(y, t)$ for any given x_i. In Fig.2 an $y - t$ plane and its response to upward motion are shown. Upward and downward energy for a given plane $y - t$ is shown in frames (c),(d) in Fig. 2

2. **Motion opponency.** To extract the dominant motion, within a considered time frame, and to discard inconsistent or flickering motion information, regarded as not salient, opponent motion is computed. As shown in [15] indeed, only coherent motion has a distinct signature resulting from a significant amount of motion opponent energy. This must be further normalized to be made comparable with other oriented energies. Here we show a measure of horizontal and vertical salience for every element in the volume, obtained by:

$$
\begin{aligned}
E_h(x, y, t) &= \bigcup_y \frac{|ER_y(x, t) - EL_y(x, t)|}{ER_y(x, t) + EL_y(x, t) + \epsilon} \\
E_v(x, y, t) &= \bigcup_x \frac{|EU_x(y, t) - ED_x(y, t)|}{EU_x(y, t) + ED_x(y, t) + \epsilon}
\end{aligned}
\tag{4}
$$

Here ϵ is a constant avoiding dividing by 0. Horizontal energy will have negative values as leftward motion prevails, and positive values as rightward motion prevails. Correspondingly, the same behaviour holds for vertical energy and downward and upward motion. In Fig. 3 horizontal and vertical energies summed along the temporal axis are shown. As it can be noted, the faster dot has a bigger overall amount of both energies as it has described more than a whole orbit, while other dots have not completed their orbits. In Fig.4 3D plots are shown, comparing horizontal energy variations between the faster dot and one of the other dots.

Finally we low-pass filter obtained energies by a 5-binomial tap to remove high frequencies.

3 Saliency Map Construction

At this point we have obtained a measure of two motion features and we have to combine them together to form a sole measure of motion salience. In [16] salience of each spatio-temporal point is given by selection of the maximal value between E_h and E_v. We consider both channels as relevant to salience definition and compose them together as components of the total energy. Nevertheless we want our map to contain only most relevant points, hence we perform first a local maxima and minima selection (corresponding to high rightward/upward energy and leftward/downward energy, respectively), then we compute the total amount of energy for each point by combining the two oriented energies, and finally we produce an overall saliency map for the considered volume.

We can, thus, distinguish coherent motion by finding the local maxima and the local minima of the above defined oriented energies. This can be obtained by closed formulae by taking the derivative and equating it to zero. The local extrema (i.e. a local maximum or local minimum) can be easily identified as follows. Let

$$A_\alpha = \begin{bmatrix} \alpha_1^o & \alpha_1^e \\ -\alpha_2^o & -\alpha_2^e \end{bmatrix} \quad B_\alpha = \begin{bmatrix} \alpha_1^o & \alpha_1^e \\ \alpha_2^o & \alpha_2^e \end{bmatrix} \tag{5}$$

and

$$M_G = \begin{bmatrix} G_{\theta_1}^{o'}(x,t) * I(x,t) & G_{\theta_2}^{o'}(x,t) * I(x,t) \\ G_{\theta_1}^{e'}(x,t) * I(x,t) & G_{\theta_2}^{e'}(x,t) * I(x,t) \end{bmatrix} \tag{6}$$

here $\alpha_i^w = 2G_{\theta_i}^w(x,t) * I(x,t)$, with $i = 1, 2$ and $w \in \{o, e\}$. On the other hand $G_{\theta_i}^{w'}$, with $i = 1, 2$ and $w \in \{o, e\}$ is the derivative of the Gabor filter used (see 1), w.r.t. t, namely:

$$G_{\theta_i}^{w'}(x,t) = \left(-\frac{t}{2\sigma_t^2}\cos(\omega_x x + \omega_t t) - \left(-\frac{x^2}{2\sigma_x^2} - \frac{t^2}{2\sigma_t^2}\sin(\omega_x x + \omega_t t)\omega_t \right) \right)$$
$$exp\left(\left(-\frac{x^2}{2\sigma_x^2} - \frac{t^2}{2\sigma_t^2} \right)\cos(\omega_x x + \omega_t t) \right) \tag{7}$$

Local maxima for E_h and E_v are thus:

$$max(E) = \frac{trace(A_\alpha M_G)}{trace(B_\alpha M_G)} \tag{8}$$

Once we have found local maxima in E_h and E_v, we assign to remaining locations null energy. We define a global energy measure by taking for each point the norm of the vector given by the two components:

$$E(x,y,t) = \sqrt{E_h(x,y,t)^2 + E_v(x,y,t)^2} \tag{9}$$

We have now achieved a measure of the total energy developed in each point of the spatiotemporal volume.

Subsequently to enhance in every frame locations that have displayed relevant motion and substantial contrast with respect to the surroundings, a center-surround mechanism is applied. Gaussian pyramids are built on each $x - y$ plane of the energy volume and on-center differences are computed on each layer, by subtracting to each pixel the mean of its surroundings. Then across-scale addiction leads to a single conspicuity map. On-off differences were computed straightforwardly by means of integral images. Conspicuity maps (CM) were weighted dividing by the number of elements above a certain threshold. We refer to [17] for details.

These maps do not take into account what has happened before, thus highlighting local motion energy only with respect to a precise time step. That is to say, they do not have memory. To integrate the observed energy over a given time slot, we compute for each $x - y$ frame a 2D saliency map SM by summing previously accumulated energy along the temporal direction:

$$SM(x,y,t) = \int_0^t CM(x,y,\tau)d\tau \tag{10}$$

This is consistent with the intuition that our visual system integrates in continuous time perceived motion (visual persistence). In this way for each spatiotemporal volume processed we find a final SM summing up relevant motion information for a given time

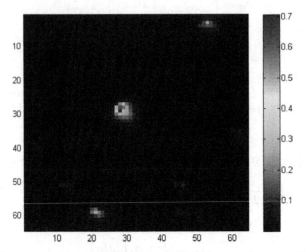

Fig. 5. Final saliency map of the example sequence in Fig. 1 for a spatiotemporal interval of about 2 sec

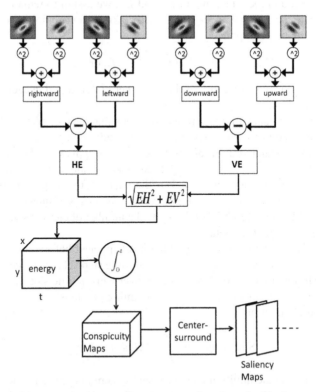

Fig. 6. The figure illustrates the framework main processing, from the filtering stage (via odd and even Gabor filters) to energy computation and saliency map construction

unit. Just these last SMs could be saved at the end of every time unit and used for successive processing and broader analysis. Flickering or incoherent motion are indeed cancelled in the final map, as illustrated in the next section 4.

In Fig.5 the final SM, computed over a $\sim 2sec$. time slot, is presented. The most significant location corresponds to the faster dot. Summing up, the main operations and data processing of the system are depicted in fig.6. The first layer performs Gabor filtering along diagonal directions, the second sums squared outputs of quadrature pairs, lowpass filters them and combines them in a single energy measure. Finally via center-surround and integration saliency maps are produced every p frames.

4 Experiments

We have tested the described framework under different motion conditions. The results shown below confirm the relevance of consistent motion, in contrast with other random motions, which our framework has been able to capture. First we tried it on synthetic sequences displaying pop-out stimuli of coherent motion. In fig.7, (a), a multitude of

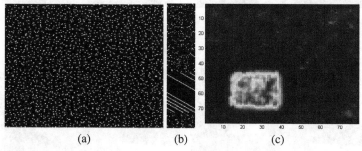

(a)   (b)   (c)

Fig. 7. Example of coherent motion pop-out: (a) original stimulus, a square of vertical moving dots is present bottom left, while the rest of the image displays flickering dots (316x316x40 frames). (b) A y-t plane (x=110) showing the tracks left by the dots in the square. (c) Saliency map.

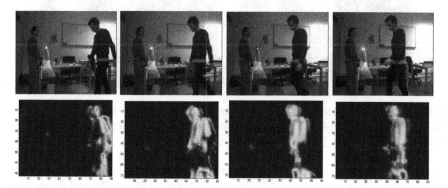

Fig. 8. Test results for a video sequence (15 fps) displaying a person walking, while the lamp is switched on and off, generating random scintillation. Some frames (t=5,10,15,20) are shown with their respective incremental salience maps.

Fig. 9. Test results for a video sequence (15 fps) displaying two people shaking hands. Eye fixations of the observer are displayed with a green dot. Frames are shown with respective incremental salience maps.

flickering dots is presented [2]. Just a square of dots bottom left moves coherently (vertically), as shown by tracks in one of the corresponding $y - t$ planes (b), and as correctly highlighted by the corresponding saliency map, (c).

---

[2] Video: http://www.opticsinfobase.org/viewmedia.cfm?uri=oe-11-13-1577&seq=1

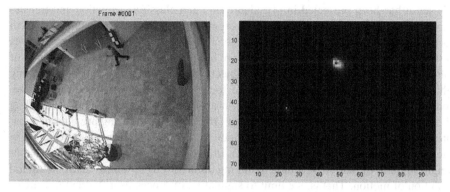

Fig. 10. Left: example frame from a surveillance camera sequence. Right: corresponding saliency map.

Apart from synthetic stimuli, used as testbed, we applied our system to real world scenes as well. For example with people moving while some scintillation is disturbing, or with different contemporary motions, such as people meeting and shaking hands. In Fig. 8 we show incremental building of the saliency map for a time lag of 5 frames (at a frame rate of $\sim$ 15 fps) in the case of a person walking, while another person is switching on and off a lamp with an intermittence of $0.3sec.$. Locations gaining most coherent energy become more and more salient.

On the other hand, we have tested the framework also in comparison with human attention. In this setting we are highlighting only the motion component of saliency, and yet the integrated saliency (static and dynamic) is not represented, but it is interesting to notice how the experiments emphasise the convolution of composite motion.

Fig.9 illustrates a video sequence showing the contemporary motion of two persons. The video has been collected by a device that can project the current eye gaze. While the system is pretty much bottom-up, one should consider that human attention is influenced by the Theory of Mind, i.e. in a top-down manner, when observing other human beings, we try to infer their mental states and thus we mostly look at their faces. Nevertheless from the experiment emerges an interesting overlapping of human attention with some of the locations highlighted in the SM, shown in the Figure. Being the movement more composite a $1sec$ lag volume was analysed as basic unit. A further study should investigate relations between temporal scale, processed time range and the motion observed in order to achieve more meaningful saliency maps.

Finally, we show one last application to a surveillance scenario in fig.10[3].

5 Conclusions

In this paper we presented a method to address the issue of attending to motion in the framework of an attentive architecture. An exhaustive and complex architecture in

---

[3] Video sequences and saliency maps available on
http://www.dis.uniroma1.it/~belardinelli/video.htm

this sense has been presented in [4], identifying and classifying rotation and translation motion patterns in a hierarchical process embedding velocity gradient. That system goes deeply into neural representation of the motion hierarchy, modelling sensitivities in the different areas and tackling the feature binding problem to achieve grouping of coherently moving features into a single object. Hence selection is helped by a strong inhibition process on the spatial level. We focussed primarily on the selective functionality of attention which can modulate in a top-down way salience and attending of certain motion direction. As explained in [18], motion is a strong pre-attentive feature eliciting attention and conversely attention to relevant motion can help selection in tracking tasks with moving distractors. We do not classify types of motion but rather weight salience of motion direction, inhibiting uninteresting or uninformative temporal evolution of motion. That is, we limit to simulation of processes in V1 and MT areas. Our approach is based on the extraction of coherent motion information in the form of energy along some preferred directions in the 2D planes defined by one spatial dimension and the temporal dimension. Spatiotemporal filtering for extraction of edges related to oriented energy has been shown in the literature as psychophysiologically plausible to model motion perception. We use this process to obtain feature 3D maps related to vertical and horizontal motion and combine them to form a unique energy map. Single conspicuity maps for every frame are formed by means of center-surround mechanism, and summed along the temporal axis to get a final salience map.

Interesting issues arise regarding composition and selection of different spatiotemporal bands with respect to specific motion events and patterns. These will be tackled in a further study. Further tests and comparisons with human attention will also provide indications on top-down biasing of motion salience.

References

1. Tsotsos, J., Culhane, S., Wai, W., Lai, Y., Davis, N., Nuflo, F.: Modeling visual attention via selective tuning. Artifical Intelligence 78, 507–547 (1995)
2. Itti, L., Koch, C.: Computational modeling of visual attention. Nature Reviews Neuroscience 2(3), 194–203 (2001)
3. Frintrop, S.: VOCUS: A Visual Attention System for Object Detection and Goal-Directed Search. LNCS (LNAI), vol. 3899. Springer, Heidelberg (2006)
4. Tsotsos, J.K., Liu, Y., Martinez-Trujillo, J.C., Pomplun, M., Simine, E., Zhou, K.: Attending to visual motion. Comput. Vis. Image Underst. 100(1-2), 3–40 (2005)
5. Adelson, E.H., Bergen, J.R.: Spatiotemporal energy models for the perception of motion. J. of the Optical Society of America A 2(2), 284–299 (1985)
6. Heeger, D.J.: Model for the extraction of image flow. Journal of the Optical Society of America A: Optics, Image Science, and Vision 4(8), 1455–1471 (1987)
7. Watson, A.B., Ahumada, A.J.J.: Model of human visual-motion sensing. Journal of the Optical Society of America A: Optics, Image Science, and Vision 2(2), 322–342 (1987)
8. Simoncelli, E.: Local analysis of visual motion. In: Chalupa, L.M., Werner, J.S. (eds.) The Visual Neuroscience, pp. 1616–1623. MIT Press, Cambridge (2003)
9. Hillstrom, A.P., Yantis, S.: Visual motion and attentional capture. Perception & Psychophysics 55, 399–411 (1994)
10. Rosenholz, R.: A simple saliency model predicts a number of motion popout phenomena. Vision Research 39, 3157–3163 (1999)

11. Singh, V.K., Maji, S., Mukerjee, A.: Confidence based updation of motion conspicuity in dynamic scenes. In: Proceedings of the The 3rd Canadian Conference on Computer and Robot Vision (CRV 2006), p. 13 (2006)
12. Milanese, R., Gil, S., Pun, T.: Attentive mechanisms for dynamic and static scene analysis. Optical Engineering 34, 2428–2434 (1995)
13. Viola, P., Jones, M.J., Snow, D.: Detecting pedestrians using patterns of motion and appearance. In: ICCV 2003: Proceedings of the Ninth IEEE International Conference on Computer Vision, p. 734 (2003)
14. Kwak, S., Ko, B., Byun, H.: Salient human detection for robot vision. Pattern Anal. Appl. 10(4), 291–299 (2007)
15. Wildes, R.P., Bergen, J.R.: Qualitative spatiotemporal analysis using an oriented energy representation. In: Vernon, D. (ed.) ECCV 2000, Part II. LNCS, vol. 1843, pp. 768–784. Springer, Heidelberg (2000)
16. Wildes, R.P.: A measure of motion salience for surveillance applications. In: International Conference on Image Processing (ICIP 1998), pp. 183–187 (1998)
17. Frintrop, S., Klodt, M., Rome, E.: A real-time visual attention system using integral images. In: Proc. of the 5th International Conference on Computer Vision Systems (ICVS 2007) (2007)
18. Cavanagh, P.: Attention-based motion perception. Science 257(5076), 1563–1565 (1992)

Model Based Analysis of fMRI-Data: Applying the sSoTS Framework to the Neural Basic of Preview Search

Eirini Mavritsaki, Harriet Allen, and Glyn Humphreys

Behavioural Brain Sciences Centre, School of Psychology,
University of Birmingham, Edgbaston, B15 2TT, UK

Abstract. The current work aims to unveil the neural circuits underlying visual search over time and space by using a model-based analysis
of behavioural and fMRI data. It has been suggested by Watson and
Humphreys [31] that the prioritization of new stimuli presented in our
visual field can be helped by the active ignoring of old items, a process
they termed visual marking. Studies using fMRI link the marking process
with activation in superior parietal areas and the precuneus [4,18,27,26].
Marking has been simulated previously using a neural-level account of
search, the spiking Search over Time and Space (sSoTS) model, which
incorporates inhibitory as well as excitatory mechanisms to guide visual
selection. Here we used sSoTS to help decompose the fMRI signals found
in a preview search procedure, when participants search for a new target whilst ignoring old distractors. The time course of activity linked to
inhibitory and excitatory processes in the model was used as a regressor
for the fMRI data. The results showed that different neural networks
were correlated with top-down excitation and top-down inhibition in the
model, enabling us to fractionate brain regions previously linked to visual marking. We discuss the contribution of model-based analysis for
decomposing fMRI data.

1 Introduction

1.1 Human Visual Search over Space and Time

The visual world contains a vast amount of information, only some of which is
relevant to our behaviour. It is therefore essential to employ selection processes
to enable us to separate relevant from non-relevant information. In order to
understand both the functional mechanisms of selection, and the underlying
neural substrates, investigators are increasingly combining behavioural studies
with fMRI analyses which reflect functional activity in different brain regions as
selection takes place. However, given the limited spatial and temporal resolution
of fMRI, it is often difficult to separate the different functional processes that
may contribute to visual selection. Moreover different functional processes can
combine to influence selection. For instance, selection may be contingent on both
excitatory processes that guide attention to a target and on inhibitory processes

L. Paletta and J.K. Tsotsos (Eds.): WAPCV 2008, LNAI 5395, pp. 124–138, 2009.

that guide attention away from distractors [5]. In such cases, fMRI contrasts between (say) easy and difficult search operations fail to distinguish the different functional operations involved. One way to advance the functional analysis of fMRI data in such cases is to link the data to an explicit model of performance, which does distinguish between the different functional processes, and which can be used to predict the variation in fMRI signal as the different processes take place. Here we present an example of this using the spiking Search over Time and Space (sSoTS) model of visual search [23]. We show how sSoTS can be used to pull-apart fMRI signals associated with excitatory and inhibitory processes in search, providing a more detailed analysis of the relations between cognitive and neuronal function.

It is well known that humans utilize "selection by space" to process only information at certain locations. However, only recently have studies been designed to examine how temporal cues can be used to guide visual selection. Traditionally, in visual search tasks participants are asked to find a known target item amongst irrelevant distractor items, and the time it takes participant to identify the target is measured (the reaction time (RT)). The slope of the search function (RT relative to the display size of distractors) depends on the spatial features of the target and distractor items. Watson and Humphreys [31] devised a new version of visual search where the temporal as well as the spatial features of targets and distractors were varied. They adapted a standard colour-form conjunction task, but presented half of the distractors (the preview) prior to the other distractors and the target (when present). They showed that this preview search condition was facilitated relative to the standard conjunction search, with search efficiency approximating that found when the new items were presented alone (the 'single feature baseline'). Watson and Humphreys [31] proposed that temporal prioritisation in search tasks depends, at least in part, on the active ignoring of old items – a process they termed visual marking. Humphreys et al. [17] showed that visual marking is disrupted when a secondary task must be conducted during the preview, consistent with the secondary task disrupting top-down ignoring of old items. In addition this, there is also evidence for top-down excitatory biases influencing search. For example a positive bias for expected target properties can offset the effects of an inhibitory bias against the features of old distractors [16] (induced by, for example, instructions or changes in display).

There is now considerable evidence that search is contingent on a network of neural circuits in frontal and parietal cortex that control both voluntary and reflexive orienting of attention to visual information [7]. These neural regions also overlap with areas involved in selecting targets on the basis of their temporal properties [20,8], suggesting that common neural processes may mediate search not only across space but also across time – as when we prioritise the selection of new over old stimuli. The inter-play between the different parts of this fronto-parietal circuit however remains much less understood.

There have now been several brain imaging studies of preview search [27,26] which converge in demonstrating that the preview period is associated with activation within the superior parietal cortex and the precuneus. Allen et al. [4]

examined preview search both when a preview task was carried out alone and under conditions of secondary task load (a visual memory task was interleaved with preview search). In a single feature baseline, the participant had to locate a blue house target amongst red house distractors. In a conjunction condition, the same target had to be found amongst blue faces and red house distractors. In the preview condition, the preview items (blue faces) appeared 2 sec before the search display (red houses and blue house target). In the visual memory task participants had to memorise the positions of dots presented before the preview display. The, after the presentation of the preview, either the dots re-appeared or the search display was presented. When the dots re-appeared the task was to judge whether one had moved location. When the search display appeared the task was to locate the target (left or right of screen). Figure 1 shows the different conditions. This study used faces and houses as search items rather that the typical lines or letters. This allowed Allen et al. [4] to draw conclusions about the activity in stimulus-specific cortex (e.f. fusiform face area). Although there are differences in behaviour with these more complex stimuli, crucially, Allen et al. found a behavioural advantage for preview search which decreased when there was a memory load. Active ignoring of the preview display was associated with activation in a network of brain areas in posterior parietal cortex. These same regions were active during the visual memory task and decreased their activation for preview displays when the memory task was imposed.

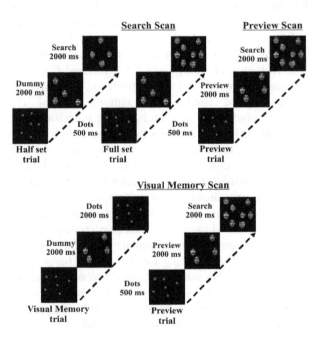

Fig. 1. This figure presents the displays for the 3 conditions in Allen et al. [4], for the three scans (Search scan, Preview scan, and Working Memory scan)

1.2   Modelling Search

Over the past ten years, increasingly sophisticated computational models of visual search and selection have been proposed [15,9,19,24]. The importance of these models is that they generate a system-level account of performance, emerging from interactions between different local components. This provides a means of examining how interactions within a complex network generate coherent behaviour.

The majority of models to-date have used relatively high-level connectionist architectures, where (e.g.) activity within any processing unit typically mimics the behaviour of many hundreds rather than individual neurons [see [15] for an example]. Such models not only operate at a level of abstraction across individual neurons (operating at a 'mean field' levels; see [15]), but they also very often include network properties divorced from real neuronal structures (e.g., with units being both excitatory and inhibitory, depending on the sign of their connection to other units). One exception to this approach comes from the work of Deco and colleagues [9,11] who have simulated aspects of human attention with models based on 'integrate and fire' neurons. These networks utilise biologically plausible activation functions and generate outputs in terms of neuronal spikes (rather than, e.g., a continuous value, as in many connectionist systems). Deco and colleagues have shown how classic 'attentional' (serial) aspects of human search can be simulated by such models even when the models have a purely parallel processing architecture. This provides an existence proof that a model incorporating details of neuronal activation functions can capture aspects of human visual attention.

One attempt to simulate human search over time as well as space has been made using the spiking Search over Time and Space model (sSoTS) [22,23], which represents an extension of the original work of Deco and Zihl [9]. sSoTS uses a system of spiking neurons modulated by NMDA, AMPA, GABA transmitters along with an IAHP current, as originally presented by Deco and Rolls [27,26] (see also Brunel and Wang [6]). sSoTS is separated into processing units that encode the presence of independently coded features (e.g. colour and form) (see Figure 2). The feature maps can be thought as high-level representations for groups of low level of features. There is in addition a 'location map' in which units respond to the presence of any feature at a given position. At each location (in the feature maps and the location map), there is a pool of spiking neurons, providing some redundancy in the coding of visual information. The feature maps may correspond to collections of neurons in the posterior ventral cortex (e.g., V4), while the location map may correspond to collections of neurons in dorsal (posterior parietal) cortex. There are inhibitory interactions across different pools in the feature maps, representing a form of lateral inhibition between like elements. There are also inhibitory interactions between pools corresponding to the same location in different feature maps in the same feature domain (e.g., between blue and green units, but not between blue and H units), so that a given location will tend to support only one feature value within a domain. Search for a specific item is simulated by giving additional activity into the feature maps

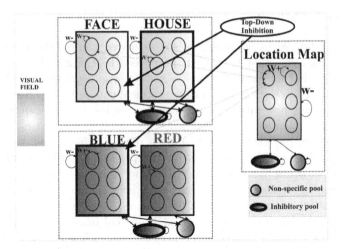

Fig. 2. The architecture of the sSoTS model: The maps outlined in bold (Blue and House maps) receive top-down excitation (for the expected target) and the maps linked to the external inhibitory pool (the Blue and Face maps) receive the top-down inhibition (for the features of the preview)

corresponding to the properties of the target; this corresponds to an expectation of the target. This activity combines with activity from the stimuli presented in the search display, and the output from each pool of neurons in each feature map is fed-forward into the map of locations. Activity in the location map provides an index of 'saliency' irrespective of the feature values involved (cf.[19]), since the location units represent the strength of evidence for 'something' occupying each position, but they are 'blind' to the features present (which are summed across the feature maps). There is then also feedback activation from the pool of units corresponding to each position in the map of locations to units at the corresponding location in the feature maps, supporting the selection of features that are linked to the highest saliency value. Over time, the model converges upon a target, with reaction times (RTs) based on the real-time operation of the neurons.

Search efficiency in sSoTS is determined by the degree of overlap between the features of the target and those of distractors, with RTs lengthening as overlap increases and competition for selection increases. Consequently, search for a conjunction target (having no unique feature and sharing one feature with each of two distractors) is more difficult than search for a feature-defined target (differing from the distractors by a unique feature). Mavritsaki et al. [22,23] showed that search in the conjunction condition also increased linearly as a function of the display size, mimicking 'serial' search.

In addition to modelling spatial aspects of search, sSoSTs also successfully simulated data on human search over time, in the preview search paradigm [1, 20]. Provided the interval between the initial items and the search display is over 450ms or so, the first distractors in preview search have little impact on behavioural performance [31,32]. The sSoTS model generated efficient preview search when there was an interval of over 500ms between the initial preview and the final search display. sSoTS mimics the behavioural time course due to the contribution of two processes: (i) a spike frequency-adaptation mechanism generated from a slow Ca^{2+} -activated K^+ current, which reduces the probability of spiking after an input has activated a neuron for a prolonged period [21], and (ii) a top-down inhibitory input that forms an active bias against known distractors. The slow action of frequency-adaptation simulates the time course of preview search. The top-down inhibitory bias matches data from human psychophysical studies where the detection of probes has been shown to be impaired when they fall at the locations of old, ignored distractors [3,4]. In addition, in explorations of the parameter space for sSoTS, Mavritsaki et al. [22,23] found that active inhibition was necessary to approximate the behavioural data on preview search. These results, using the sSoTS model, indicate that processes of co-operation and competition between processing units may not be sufficient to account for the full range of data on human selective attention and that factors such as frequency adaptation are required in order to simulate the temporal dynamics of visual attention.

1.3   Linking the Model to fMRI

As we have noted, imaging studies have shown a network of regions in posterior parietal cortex (PPC) (including superior parietal cortex and precuneus, extending into occipital cortex) associated with successful prioritisation of the new target and successful ignoring of the old distractors. However, the increased activation in these regions found in preview search is inherently ambiguous, because preview search is influenced by both positive expectancies for targets and inhibitory suppression of distractors [5]. This ambiguity is not apparent in the sSoTS model, though, where effects of top-down expectancies and inhibitory biases against distractors can be distinguished. For example, the map associated with the feature of the old distractors that does not re-occur in the search display (i.e., the map for face stimuli, in the experiment of Allen et al. [4]) uniquely receives top-down inhibition in sSoTS. The map corresponding to the feature of the target not present in the old distractors (i.e. houses in Allen et al. [4]) uniquely receives top-down activation. The changes in activity over time in these maps may be used to predict changes in the fMRI signal linked, respectively, to top-down expectancies and inhibition in preview search. The distinct time courses of activation in the model may then be used to pull-apart activity from within the regions linked to preview search, allowing us to isolate the neural regions concerned with excitatory and inhibitory modulation of processing. We report an analysis of fMRI data on preview search taking this approach.

2   The Architecture of sSoTs

sSoTS consists of spiking neurons organised into pools containing a number of units with similar biophysical properties and inputs. The simulations were based on a highly simplified case where there were six positions in the visual field, allowing up to 6 items in the final search displays. sSoTS has three layers of retinotopically-organised units, each containing neurons that are activated on the basis of a stimulus falling at the appropriate spatial position. There is one layer for each feature dimension ("colour" and "form") and one layer for the location map (Figure 2). The feature maps encode information related to the features of the items presented in an experiment – in this case, Allen et al. [4]. For Allen et al. [4], the two different features encoded are colour and object shape, which in this case is house or face. Here the feature dimension "colour" encoded information on the basis of whether a blue or red colour was presented in the visual field at a given position i, (i=1,...,6) (creating activity in the red and blue feature maps). The feature dimension "form" encoded information on the basis of whether there is a house or face present in the visual field at a given position i (i=1,..,6). The pools in the location map sum activity from the different feature maps to represent the overall activity for the corresponding positions in the visual field. Each of the layers contains one inhibitory pool (see also [11]) and one non-specific pool, along with the feature maps.

The system used and the connections are illustrated in Figure 2. More details about the architecture of sSoTs and the organisation of the units (neurons) in the network can be found in Mavritsaki et al. [22].

The units in the model are integrate-and-fire neurons with sub-threshold given by the equation

$$C_m \frac{dV(t)}{dt} = -g_m(V(t) - V_L) - I_{syn}(t) + I_{AHP} \qquad (1)$$

Where C_m is the membrane capacitance where different values are given for excitatory $C_m ex$ and inhibitory $C_m in$ neurons; g_m is the membrane leak conductance where different values are also given for excitatory $g_m ex$ and inhibitory $g_m in$ neurons; V_L is the resting potential; I_{syn} is the synaptic current and I_{AHP} is the current term for the frequency adaptation mechanism. The values for the above parameters as well as the threshold V_{tbr} and the reset potential can be found in Mavritsaki et al. [23] . The description of the synaptic currents used (*NMDA, GABA, AMPA*) can also be found in Mavritsaki et al. [7].

The parameters for the simulations were established in baseline conditions with 'single feature' and 'conjunction' search tasks as reported by Watson and Humphreys[31] and Allen et al. [4] (conjunction search: blue house target vs. red houses and blue faces distractors; feature search: blue face target vs. red houses distractors). The generation of efficient and less efficient (linear) search functions in these conditions replicates the results of Allen et al. [4]. These same parameters were then used to simulate preview search. The parameter w^+ represents the strength of connections between the neurons in each pool, while w^- represents the strength of connections between the pools within and across each feature map. The target also benefited from an extra top-down input λ_{att} given to those feature maps that represent the target's characteristics (i.e., the colour blue and the letter H). The presence of an object in the visual field was signified by adding an additional λ_{in} value given to the external input that the system received.

Overall the input that a pool could receive was $v_{ext} = v_{ext} + (\lambda_{in} + \lambda_{att})/N_{ext}$. In preview search top-down attention (λ_{att}) was applied to the target's feature maps at the onset of the search display.

RTs were based on the time taken for the firing rate of the pool in the location map to cross a relative threshold (thr). If the selected pool corresponded to the target then the search was successful (a hit trial). If the pool that crossed the threshold corresponded to a distractor rather than the target then the target was 'missed'. Note, however, that if the parameters were set so that the target's pool was the winner on every trial, only small differences in the slopes were observed between conjunction and single feature search, due to target activation saturating the system. Accordingly, search was run under conditions in which some errors occurred, mimicking human data. Detailed simulations, were run at the spiking level only, to match the experimental results [4]. Additionally, to simulate the working memory effect, we reduced slightly the top-down inhibition during the 'working memory' trials – assuming this is equivalent to the effects generated when human participants hold another stimulus in working memory during the preview period.

3   Applying the sSoTS Model to fMRI Data

3.1   Extraction of Activation Maps for Top Down Inhibition and Excitation

During the preview period activation in the model is affected by several factors: top-down excitation (for the target), top-down inhibition (for old distractors) and passive inhibition caused by frequency adaptation. In order to be able to compare the fMRI data with the activation patterns in the model we extracted activation maps from the model related to the above mechanisms. For example, consider preview search for a new blue house target amongst previewed blue faces and new red houses distractors (see[4]). In sSoTS there is a positive bias applied to maps representing the features of targets, for Allen et al. [4] the target is the blue house, therefore the map that encodes the shape "house" and the map that encodes the colour "blue" receive top-down excitation. Furthermore, there is an inhibitory bias applied to maps representing the features of old distractors (distractors presented before the presentation of the search display), these distractors are blue faces, so the map that encodes the shape "face" and the map that encodes the colour "blue" both receive top-down inhibition. By tracing activity in the house, face and blue maps, we can correlate brain activity with active excitatory and inhibitory biases in the model. Note that we are interested in activity relating to these biases and processes, not to the distractor features or colours.

To extract the brain activity relating to these processes, we first extracted a time course of the activity in each of the sSoTS maps (2 x shape, 2 x feature and the location map) over the experiment of Allen et al. [4]. These time courses were convolved with a standard estimate of the heamodynamic function and used as regressors for the fMRI activity (see below). To estimate the activations

Table 1. Map Extraction

Single Feature and Conjunction Map Extraction		Preview Map Extraction	
Maps	Positive Bias SF and CJ	Inhibitory Bias PV	Positive Bias PV
Face	NO	YES	NO
House	YES	NO	YES
Blue	YES	YES	YES
Red	NO	NO	NO

associated with positive biases for targets and inhibitory biases against distractors (see Table 1) we compared the activations found for each map (for both the conjunction and preview search conditions). Thus, for conjunction search, the positive top-down bias was given by:

$$\textbf{(Target form – distractor form)}$$
$$\textbf{+}$$
$$\textbf{(Target colour – Distractor Colour)}$$

i.e: (House – Face+ Blue – Red)

For preview search the top-down excitation was given by:

$$\textbf{(Map with only Positive Bias –Map with no bias)}$$
$$\textbf{+}$$
$$\textbf{(Map with Positive and Negative Bias – Map with only Negative Bias)}$$

i.e: (House–Red+Blue–Face)

For preview search the top-down inhibition was given by:

$$\textbf{(Map with only Negative Bias – Map with no Bias)}$$
$$\textbf{+}$$
$$\textbf{(Map with Positive and Negative Bias – Map with only Positive Bias)}$$

i.e: (Face–Red+Blue–House)

3.2   Comparison of fMRI Data with Model Bold Responses

Activation in sSoTS was linked to the human fMRI data by taking into account the delay that is present in the fMRI bold signal (about 5-9 sec) [12]. To do this, activity in the model was convolved with a haemodynamic response function [13,14,10]. Previous work by Gorchs and Deco [13] simulated the bold response by taking the average pool activity in a given location in the model and convolved this with a Poisson distribution. The result from the convolution was then compared with bold responses taken from the fMRI data (from the corresponding simulated region). Furthermore, instead of using the average pool activity the synaptic activity can also be employed. Deco et al. [10] used the synaptic activity from his model and convolved it with the haemodynamic response function suggested by Glover [14]. In our effort to compare our theoretical data with the fMRI experimental data we used the average synaptic activity from the pools in the model's feature maps. This average synaptic pool activity was then directly compared with the observed bold data from Allen et al. [4], using the synaptic activity as regressors for the fMRI analysis.

We note that there was no top-down inhibitory bias applied during conjunction search. However, activity in the same maps was examined in order to provide a baseline with the preview search task. After extracting the activity maps from the model, we averaged over 20 trials for each condition and we took the changing time course of activity reflecting top-down inhibition and top-down excitation activity for each condition. This activity was convolved using an assumed haemodynamic response function [10] to create a time series of predicted bold activity. This time series was then used as a regressor for the fMRI data in the contrasting search conditions.

fMRI analysis was done using FEAT, part of fsl (www.fmrib.ox.ac.uk/fsl). The data were pre-processed as in Allen et al. [4], including correction for head movement, within scan signal intensity normalisation, high pass temporal filtering (to remove slow wave artifacts). The time course for each map in the model was entered as a separate regressor. Positive and negative biases were estimated by combining the regressors for each map as desribed above. Z (Gaussianised T/F) statistic image were thresholded using clusters determined by Z>2.3 and a (corrected) cluster significance thresholded of P=0.05.

4 Results

The behavioural results generated by sSoTS matched the classical findings on single feature, conjunction and preview search [31]. In the single feature condition (the half set baseline), the search slope was 14 ms/item; for the preview condition it was 12 ms/item, and it was 46 ms/item for the conjunction condition (the full set baseline). When a working memory task was added (the loaded search condition), the slope of the preview condition increased to 19 ms/item (see Figure 3).

We then took the time courses of activation reflecting the top-down excitatory and inhibitory activity in sSoTS's feature maps and applied these as regressors to the fMRI data associated with the preview condition reported by Allen et al. [2]. In this study we sought areas where BOLD activity was related to excitatory and inhibitory activity. Allen et al. [4] reported activation in posterior parietal cortex (superior parietal lobe and precuneus) linked to the dummy preview condition. We found a reliable correlation (p<0.001 for all correlations) in right lateral parietal cortex for top-down excitatory activity predicted by sSoTS. In contrast, top-down inhibitory activity in the model was correlated with fMRI activation in the medial precuneus (Z=50) (Figure 4). Here the model-based analysis distinguishes two functionally different operations taking place when observers attempt to ignore the preview and to prioritise search to new items [5].

We also examined the differences between bold activity in the preview and conjunction search conditions in relation to the activation differences between

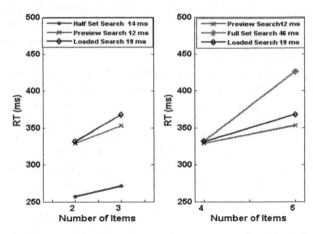

Fig. 3. The slopes generated by sSoTS for single feature search (the half set baseline), conjunction search (the full set baseline), standard preview search and preview search with a working memory load (the loaded search condition). On the left we present the preview and the loaded search conditions compared with the half set baseline, with the display size matched to the number of items in the second search display in the preview condition. On the right we show the preview and the loaded search conditions in comparison with the full set baseline, with the display size matched to the number of items on the final screen of the preview condition (preview + search items).

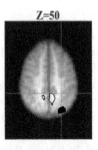

Fig. 4. This figure presents the areas that show bold activity correlated with top-down inhibition in sSoTS (white with black outline) and those where bold activity correlated with top-down excitation in the model (black with white outline). Top-down inhibition in the model (maps: (1-4)+(3-2)) was associated with activity the medial precuneus, while top-down excitation in the model (maps: (2-4)+(3-1)) was associated with activity in the lateral parietal cortex (right hemisphere).

the these conditions apparent in sSoTS (comparing activity in the critical maps in preview and conjunction search). In sSoTS these activation differences are driven by the application of top-down inhibition in preview search. The results showed a reliable correlation between the activation differences in sSoTS and increased activation in the precuneus in preview search compared with the conjunction condition. There was also a correlation between differences in activity

Z=52

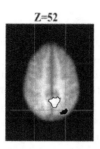

Fig. 5. Comparisons between preview and conjunction search (the full set baseline). The white with black outline regions reflect correlations between (i) top-down inhibitory activity in sSoTS and (ii) increased activation in preview compared with conjunction search. The black with white outline regions reflect correlations between (i) top-down inhibitory activity in sSoTS and (ii) greater activation in conjunction search compared with preview search.

Z=50

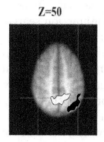

Fig. 6. Comparisons between the standard preview condition and the condition where preview search was conducted with a working memory load (the loaded search condition) [4]. The white with black outline regions reflect correlations between (i) top-down inhibitory activity in sSoTS and (ii) increased activation in standard preview search compared with the loaded search condition. The blue with white outline regions reflect correlations between (i) top-down inhibitory activity in sSoTS and (ii) increased activation in the loaded search condition compared with standard preview search.

in the conjunction and preview conditions in sSoTS and increased activity for the conjunction condition over the preview condition in lateral parietal cortex (Z=52) (see Figure 5). This may reflect the increased role of excitatory guidance to the target in the conjunction condition.

Finally, we evaluated the differences in activity between the standard preview condition and preview search conducted with a memory load. The differences in activity between these two conditions in sSoTS was correlated with (i) an increase in bold activity in the standard preview compared with the working memory condition in the precuneus, and (ii) an increase in bold activity in the working memory condition compared with the standard preview in lateral parietal cortex (Z=50) (Figure 6). These results fit with there being reduced inhibitory activity under conditions of working memory load, along with an increased role for top-down activation for the target under the more difficult working memory condition.

5   Conclusions

sSoTS replicated successfully the behavioural results from Allen et al. [4]. Activity in the model linked to top-down excitation and inhibition also correlated with the bold signal in posterior parietal cortex. Prior fMRI studies have demonstrated increased activity in posterior parietal cortex linked to preview search, but differences in excitatory and inhibitory influences have not been separated. In sSoTS the activation associated with top-down excitation and inhibition can be distinguished. We showed that bold activity in the precuneus was associated with top-down inhibition in the model, while activity in more lateral parietal areas (particularly in the right hemisphere) correlated with top-down excitation in the model. Activation in these two regions also changed across the search conditions in accord with changes in sSoTS. Higher activation in the precuneus in preview search compared with (i) conjunction search and (ii) the working memory condition was correlated with greater inhibitory activity in the model. In contrast, there was increased activity in lateral parietal cortex associated with increased activation in (i) conjunction search and (ii) the working memory condition, compared to standard preview search, linked to increased top-down excitation in sSoTS. These data suggest that top-down inhibition may play a driving role in generating efficient preview search compared with less efficient search conditions (conjunction search and preview search with a working memory load). Top-down activation, on the other hand, appears to play a greater role in inefficient search (conjunction search, preview search with a working memory load) than in efficient preview search. This may reflect the more prolonged search taking place, which enables a greater role for top-down excitation, for the target, to emerge. The analysis demonstrates that the model-based analysis can help to identify the functional role of different brain regions in search, providing a more accurate account of the neural substrates of visual selection.

It now remains for empirical studies to test the predictions arising from this modelling-fMRI study. For example, damage to inferior parietal cortex ought to mean that patients are impaired at exploiting any positive expectancy for upcoming targets, to facilitate search. In contrast, patients with damage to more medial and superior parietal regions (including the precuneus) ought to have problems in suppressing irrelevant distractors. While damage to posterior parietal cortex has been shown to disrupt preview search [25], the precise factors involved, and whether they might differ across patients, has not been explored. The analysis with sSoTS predicts that differences should emerge as finer-grained analyses of patient sub-groups is undertaken.

References

1. Abeles, A.: Corticonics. Cambridge University Press, Cambridge (1991)
2. Agter, A., Donk, M.: Prioritized selection in visual search through onset capture and color inhibition: Evidence from a probe-dot detection task. Journal of Experimental Psychology: Human Perception and Performance 31, 722–730 (2005)

3. Allen, H.A., Humphreys, G.W.: Previewing distracters reduces their effective contrast. Vision Research 47, 2992–3000 (2007)
4. Allen, H.A., Humphreys, G.W., Matthews, P.M.: A neural marker of content-specific active ignoring. JEP: HPP 34, 286–297 (2008)
5. Braithwaite, J.J., Humphreys, G.W.: Inhibition and anticipation in visual search: Evidence from effects of color foreknowledge on preview search. Perception and Psychophysics 65, 312–337 (2003)
6. Brunel, N., Wang, X.: Effects of neuromodulation in a cortical networks model of object working memory dominated by current inhibition. Journal of Computational Neuroscience 63-85, 63–85 (2001)
7. Corbetta, M., Kincade, J.M., Ollinger, J.M., McAvoy, M.P., Shulman, G.L.: Voluntary orienting is dissociated from target detection in human posterior parietal cortex. Nature Neuroscience 3, 292–297 (2002)
8. Coull, J.T.: Neural correlates of attention and arousal: Insights from electrophysiology, functional neuroimaging and psychopharmacology. Progress in Neurobiology textb 55, 343–361 (1998)
9. Deco, G., Zihl, J.: Top-down selective visual attention: A neurodynamical approach. Visual Cognition 8(1), 119–140 (2001)
10. Deco, G., Rolls, E., Horwitz, B.: Integrating fMRI and single-cell data if visual working memory. Neurocomputing 58-60, 729–737 (2004)
11. Deco, G., Rolls, E.: Neurodynamics of biased competition and cooperation for attention: a model with spiking neuron. Journal of Neurophysiology 94, 295–313 (2005)
12. Friston, K.J., Jezzard, P., Turner, R.: Analysis of functional MRI time-series. Human Brain Mapping 1, 153–171 (1994)
13. Gorchs, S., Deco, G.: Feature-based attention in human visual cortex: simulation of fMRI data. NeuroImage 21, 36–45 (2004)
14. Deconvolution of impulse response in event-related BOLD fMRI. NeuroImage 9, 419–429 (1999)
15. Heinke, D., Humphreys, G.W.: Selective attention for identification model (SAIM): Simulating visual neglect. Computer vision and image understanding 100, 172–197 (2005)
16. Humphreys, G.W., Duncan, J., Treisman, A.: Pre-frontal cortex and the neural basis of executive function. In: Miller, E.K. (ed.) Attention space and Action: Studies in cognitive neuroscience. Oxford University Press, Oxford (1998)
17. Humphreys, G.W., Jolicoeur, P., Watson, D.: Fractionating the preview benefit in search: Dual-task decomposition of visual marking by timing and modality. Journal of Experimental Psychology: Human Performance and Perception 28(3), 640–660 (2002)
18. Humphreys, G.W., Kyllinsbaek, S., Watson, D.G., Olivers, C.N.L., Law, I., Paulson, O.: Parieto-occipital areas involved in efficient filtering in search: A time course analysis of visula marking using behavioural and functional imaging procedures. Quarterly Journal of Experimental Psychology 57A, 610–635 (2004)
19. Itti, L., Koch, C.: A saliency-based search mechanism for overt and covert shifts of visual attention. Vision Research 40, 1489–1506 (2000)
20. Kanswisher, N., Wocjiulik, E.: Visual attention: insights from brain imaging. Nature Reviews Neuroscience 1, 91–100 (2000)
21. Madison, D., Nicoll, R.: Control of the repetitive discharge of rate cal pyramidal neurons in vitro. Journal of Physiology 345, 319–331 (1984)

22. Mavritsaki, E., Heinke, D., Humphreys, G.W., Deco, G.: Suppressive effects in visual search: A neurocomputational analysis of preview search. Neurocomputing 70, 1925–1931 (2007)
23. Mavritsaki, E., Heinke, D., Humphreys, G.W., Deco, G.: A computational model of visual marking using an interconnected network of spiking neurons: The spiking Search over Time and Space model (sSoTS). Journal of Physiology Paris 100, 110–124 (2006)
24. Mozer, M.C., Sitton, M.: Computational modelling of spatial attention. In: Pashler (ed.) Attention, pp. 341–388
25. Olivers, C.N.L., Humphreys, G.W.: Spatiotemporal segregation in visual search: Evidence from parietal lesions. Journal of Experimental Psychology 30(4), 667–688 (2004)
26. Olivers, C.N.L., Smith, S., Matthews, P.M., Humphreys, G.W.: Prioritizing new over old: An fMRI study of the preview search task. Human Brain Mapping 24, 69–78 (2005)
27. Pollman, S., Weidner, R., Humphreys, G.W., Olivers, C.N.L., Muller, K., Lohmann, G.: Separating segmentation and target detection in posterior parietal cortex-an event-related fMRI study of visual marking. NeuroImage 18, 310–323 (2003)
28. Rolls, E., Deco, G.: Computational neuroscience of vision. Oxford University Press, Oxford (2002)
29. Rolls, E., Treves, A.: Neural networks and brain function. Oxford University Press, Oxford (1998)
30. Tuckwell, H.: Introduction to theoretical neurobiology. Cambridge University Press, Cambridge (1998)
31. Watson, D., Humphreys, G.W.: Visual marking: Prioritizing selection fir new objects by top-down attentional inhibition of old objects. Psychological Review 104, 90–122 (1997)
32. Watson, D., Humphreys, G.W., Olivers, C.N.L.: Visual marking: using time in visual selection. Trends in Cognitive Sciences 7(4), 180–186 (2003)
33. Wilson, F., O'Scalaidhe, S., Goldman-Rakic, P.: Funcional synergism between putative gamma-aminobutyrate-containing neurons in pyramidal neurons in prefrontal cortex. Proceedings of the National Academy of Science 91, 4009–4013 (1994)

Modelling the Efficiencies and Interactions of Attentional Networks

Fehmida Hussain and Sharon Wood

Representation & Cognition Group,
Department of Informatics,
University of Sussex,
Falmer, Brighton, BN1 9QJ, UK
{f.hussain,s.wood}@sussex.ac.uk

Abstract. Posner and colleagues [38,40] assert that attention comprises three distinct anatomical areas of the brain responsible for separate aspects of attention, namely alerting, orienting and executive control. Based on this view of attention, the work presented here computationally models the attentional networks task (ANT) which can be used to assess the efficiency and interactions of these disparate networks, collectively responsible for different functions related to attention mechanisms. The present research builds upon the model of ANT to show the modulation effects of one network on the other and suggests how the model can be used to simulate neglect conditions related to attention. The model is evaluated against data sets from experimental studies and the model's fit to data is assessed statistically. Building such models of attention benefits computer vision research, as they are, well informed from both cognitive psychology and neuroscience perspectives.

1 Introduction

1.1 Theories of Attention and Attentional Networks

There are various psychological theories that try to explain how the mechanism of attention takes place. The first systematic theories of attention date back to the 1950s, describing attention as a single phenomenon based on central bottlenecks or limited processing capacity [7]. Later the focus shifted from attention in general to specific theories concerning how people chose among multiple objects, studying specific tasks. A few popular and established theories of attention are Feature Integration Theory [50], Guided Search Theory [57] Bundesen's Theory of Visual Attention [8] and the phenomenon of 'change blindness' and Coherence Theory [45].

Functional neuroimaging has enabled researchers to view many cognitive processes in the window of which brain areas are activated when various attention components are working [12,39,21,15]. There is sufficient evidence to believe that these networks can be distinguished both at cognitive and neuroanatomical levels [44]. This has led to a different kind of theory based on separate but collaborating attentional networks in which attention can be viewed as an organ system

L. Paletta and J.K. Tsotsos (Eds.): WAPCV 2008, LNAI 5395, pp. 139–152, 2009.

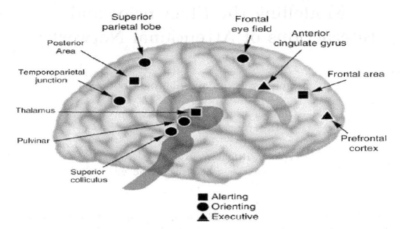

Fig. 1. The neuroanatomy of attentional networks [41, p. 6] illustrates the cortical areas involved in the three attention networks. The alerting network (squares) includes thalamic and cortical sites related to the norepinephrine system. The orienting network (circles) is centered on parietal sites and the executive network (triangles) includes the anterior cingulate and frontal areas.

or as a system of anatomical areas that consist of more specialized networks. Based on these anatomical findings, Posner proposed his three-component theory whereby attention is divided into three separate networks: namely, alertness, selectivity and processing capacity [38], later revised and renamed as alerting, orienting and executive control [39,40,41] (see Figure 1). Similarly, LaBarge's [28] triangular circuit theory of attention requires simultaneous activity of three brain regions that are connected by a triangular circuit.

Posner and colleagues state that alerting helps us to prepare for an incoming stimulus so we respond faster and more accurately. Orienting, or selective attention, helps us deal with information overload so that we can select a target among distracters in a cluttered visual scene. Finally, control helps us deal with conflicts in decision making related to attention. Although the attentional networks are anatomically and functionally independent and subtended by separate neural networks in the brain, the three networks operate under the constant influence of one another and orchestrate together to produce efficient and adaptive behavior. At first glance, it may seem that the three-component theory of attention is primarily supported from a neuroscience perspective; however, there is also support for three networks from psychophysical studies: the mechanism of orienting is in line with the classic theories of visual selective attention dealing with tasks like cueing experiments, visual search [50,57,8], and so on. The component of executive control relates to the phenomenon of cognitive control and can be supported by theories of cognitive control [47,12]. Finally, alertness provides a good explanation for theories of enhancement, giving rise to mechanisms like priming and cueing. Hence, the networks theory seems to provide a more complete view of the cognitive phenomenon of attention.

1.2 Attentional Network Test (ANT)

There are numerous tasks that have been used to study the efficiency and interactions of these attentional networks separately. For instance, alerting has been studied using a vigilance task and warning signals. Orienting has been studied using visual search tasks, spatial cueing experiments, and other visual selective attention related tasks. Finally executive control, which involves conflict resolution, is well portrayed by tasks like Stroop, Flanker, Wisconsin card sort, and so on. However, a more holistic approach would be to look at all three networks simultaneously, during execution of a single task. One such paradigm discussed below is the Attentional Network Test (ANT) developed as a behavioral measure of the efficiencies of the three attentional networks within a single task [16,46].

ANT is a computer based reaction time test which is a combination of cueing experiments [36] and a flanker task [14]. Each trial begins with a cue that informs the participant that a target is coming soon and also where it will occur. In the no-cue condition there is no signal of occurrence in time or location. The target always appears either above or below the fixation point and consists of a central arrow surrounded by flanking arrows that can either point in the same direction (congruent) or in the opposite direction (incongruent). ANT uses differences in reaction time (RT) between each experimental condition to measure the efficiency of each network. The design of ANT is illustrated in Figure 2.

The usability of ANT is very diverse and its value can be gauged from its wide application in the study of adults with borderline personality disorder [26], schizophrenia [18,55], and Alzheimer's disease [17]. Patients with attention deficits/disorders are shown to have specific deficits in the functions specifically of alerting and executive control [43,41,6]; autism has been shown to be related to the orienting network, and Alzheimer's, borderline personality disorders and schizophrenia have been shown to be related to executive control [42].

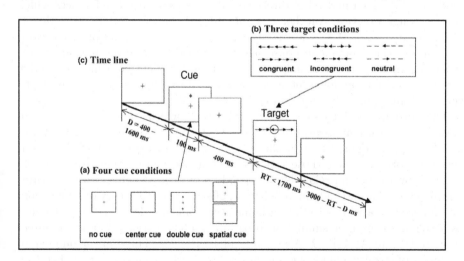

Fig. 2. A sketch depicting the design of the ANT paradigm [53, p. 121]

1.3   Computational Modeling of Attention

Computational modeling, a challenging task, is a quickly growing field in not only computer vision, but also in general in cognitive science and neuroscience. With advancements in computational modeling and progress in neuroscience, it would be insufficient to research a cognitive phenomenon from a psychology, neuroscience or computer vision perspective alone; rather, synergizing various disciplines renders tremendous benefits. There are mainly two classes of models relating to attention. There are models that emerge from the point of view of neuroscience and also neuropsychology, built to simulate the neural mechanism of the attentional processes of the brain; the objective is to be able to understand how cognitive functions like perception, memory, thinking, language, decision making, and so on arise from their neural bases. Then there is another class of models that are mainly built to solve computer vision problems. These types of models aim at building computational attention systems which have applications in the field of computer vision and robotics. Typical applications include robot navigation, surveillance tasks, industrial control, and medical imaging.

Based on these needs, there are three broad categories of modeling approaches. A popular and useful approach is that of filter based models [27,23,24] used mainly in computer vision applications. Generally this class of computational model responds to the saliency of components of the visual scene such as brightness, contrast and color, essentially corresponding to bottom-up attentional processes. The performance of such models corresponds well with psychophysical data for attention to natural scenes. Further enhancements to this approach reflect learnt associations to regularities in natural scenes, thus contributing a top-down aspect to attention [48,49,32,13]. An alternative approach to modeling uses a connectionist approach which is claimed to be more biologically plausible. A classic example of a connectionist model that simulates the Stroop task is the model of [11] which instead of direct connections uses weight differences which come through practice. Another example, SLAM (SeLective Attention Model) [35] is an extension of McClelland and Rumelhart's [31] model of visual word recognition which adds a response selection and evaluation mechanism. Selective tuning and related work [52], is a connectionist model that achieves selective tuning through a top-down hierarchy of winner-take-all processes. An in depth survey of this approach can be found in [20].

The third approach uses cognitive architectures which are mainly symbolic in nature but which may incorporate subsymbolic constructs. According to Howes and Young [22] (quoted by [19] p302), "a cognitive architecture embodies a scientific hypothesis about those aspects of human cognition that are relatively constant over time and relatively independent of task." Cognitive architectures are widely used to model human behavior, offering a broad theory of human cognition based on a wide selection of human experimental data, and implemented as a running computer simulation program [2,4,33]. Various popular architectures today are ACT-R [2] Soar [29] and EPIC [25]. There are a number of examples of cognitive models found in the literature which try to model certain aspects of attention. For example, Lovett's [30] implementation of Stroop is a good example of an ACT-R implementation of a model of cognitive control. The ACT-R theory has also been extended to include a theory of visual attention and pattern recognition whereby production rules direct attention to primitive visual features in the visual array [5]. The ACT-R theory itself embeds Posner's spotlight metaphor [36], Trieisman and Sato's feature synthesis model [51] and Wolfe's guided search model [57]. The advantage of having such a theory is two-fold: one is to model information processing limitations in obtaining information from the screen; the second is to "remove the magical degrees of freedom in going from a description of an experiment to a cognitive model." [5, p. 65].

1.4   Computational Modeling of Attentional Networks

There are various models found in the literature, such as those cited above, that are built to study a specific component of attention. However, simulating the performance of the three together has been sparse. We have come across two such models that implement the attentional networks, both simulating their performance on ANT). The first [56] is a connectionist model based on the Leabra (local error-driven and associative, biologically realistic algorithm) [34]. The second model [53] is a symbolic model based on the cognitive architecture of ACT-R 5.0. Wang and colleagues have also attempted to primitively link and compare the two approaches [54].

2   Model of Attentional Networks

The work reported in this paper is based on a reimplementation of Wang and Fan's [53] model, extending it to study the modulation effects of the attentional networks and proposing how this modeling effort can be applied in various attention related neglect conditions. It is implemented in ACT-R 6.0 [3,1] which, as described earlier, provides support for theory of visual attention [5] and incorporates both symbolic and sub-symbolic components.

2.1   Design

The model has six distinct modules which are involved in performing the generic ANT trial: fixation and cue expectation, 'cue or stimulus' processing, cue processing, stimulus expectation, stimulus processing and response. These functional components are mapped into a number of production rules within the symbolic part of the architecture that cover all the possible ANT conditions; however not all rules are fired in any one particular trial, firing depending upon the cue or stimulus.

The ACT-R model interacts with the outside world using perceptual motor modules for finding and extracting information from its *Visicon* (Visual Icon). It mimics the spotlight metaphor in which a variable size spotlight moves across a visual field, fixating on an object so that its features can be recognized. Once recognized, the object features become available for higher level processing. The implemented model uses two main buffers in the vision module: the visual-location buffer which can see the basic features but cannot recognize the semantics (as in a pre-attentive stage), and a visual buffer to which attention needs to be moved in order to do higher level processing (as in the attentive stage). The way the model deals with the visual input is a good example of the case where both pre-attentive and attentive processes work together. Capacity limits can be related to the number of items attended. In the context of ACT-R, *finsts* maintain a record of the objects that have been attended to and thus provide a mechanism which allows one to explicitly specify how many items can be attended to and for how long. *Finsts* are limited in number and how long they persist, both controlled by ACT-R parameters: the default number of *finsts* is set to four, and the default decay time is three seconds [1]. The model decides whether a stimulus is a cue or target on the basis of pre-attention, but requires full attention to process the target and respond regarding the direction of the arrow. This is in line with ACT-R's theory of attention, whereby, in order for it to know what is in the environment; it must move its attentional focus over the visual scene. It is interesting to note here that ACT-R has the ability to prevent the system from returning to previously attended objects, thus implementing the phenomenon of 'inhibition of return'. The model achieves this by allowing only items tagged as 'attended new' to be 'stuffed' into the visual-location buffer. Buffer stuffing is a mechanism in the ACT-R architecture that corresponds to the concept of bottom-up processing in visual attention. However, based on the goals of the model, the buffer is 'stuffed' using certain predefined criteria and hence reflects top-down control.

The subsymbolic part of ACT-R is used in the model to implement various parameters like rule firing time, noise, to induce randomness, utility values set to deal with conflicting productions in case of incongruency, and so on. In the case of multiple choices of matching productions, the internal conflict resolution mechanism of ACT-R is applied. In ACT-R, the utility module provides support for the productions' subsymbolic utility value which is used in conflict. This value is a numeric quantity associated with each production that can be learned while the model runs or specified in advance for each production. If there are a number of productions competing with expected utility value Uj then the probability of choosing production i is described by formula (1).

$$Probability(i) = \frac{eU_i\sqrt{2s}}{\sum_j eU_j\sqrt{2s}} \qquad (1)$$

In this default ACT-R formula [1], the summation is over all productions which are currently able to fire; s is the expected gain noise, that is the noise added to the utility values, and e is the exponential function.

2.2   Results

The model is treated as a simulated human subject in an ANT experiment, using the same dataset as used in the human studies [16], and interacting with the same experimental software [5]. The time from the stimulus presentation to the key press is recorded as the reaction time (RT). The efficiency of each network is measured using formulae (2)–(4).

$$Alerting\ efficiency\ =\ RT(no\text{-}cue) - RT(double\text{-}cue) \quad (2)$$

$$Orienting\ efficiency\ =\ RT(center\text{-}cue) - RT(spatial\text{-}cue) \quad (3)$$

$$Executive\ \ Control\ \ efficiency\ =\ RT(incongruent) - RT(congruent) \quad (4)$$

Table 1 reports the results produced by the new implementation, comparing these results with the human data and with Wang et al's earlier implementation [16,53] indicating a faithful reimplementation of the original ACT-R 5.0 model, as well as reproducing a close approximation to the original human data set.

Table 1. Comparison of Results of Fan et al's [16] Study, Wang et al's ACT-R 5.0 model [53] and the ACT-R 6.0 model presented here

	Human data			Wang's Model			ACT-R 6 new model		
	neutral	congruent	incongruent	Neutral	congruent	incongruent	neutral	congruent	incong
Nocue	529	530	605	527	526	621	520	521	592
Center	483	490	585	487	486	580	482	483	557
Double	472	479	574	467	466	562	464	459	531
spatial	442	446	515	441	441	522	441	441	527
Correlation Coefficients with human data				**0.99**			**0.97**		

3   Modelling Attention Related Disorders

As mentioned earlier, ANT has been widely used to assess which attentional networks are affected by various attention related deficits [26,55,17,43,41,6]. ANT is considered a relatively sensitive tool for assessing attention related disorders because it can closely determine the efficiency of individual attentional networks corresponding to distinct areas in the brain and can be used to assess which particular network is affected by a particular condition.

3.1   Design

The model described in Section 2 has been modified to simulate one such study which uses a modified version of ANT to assess the role of the various attentional networks in Alzheimer's disease. The study models the findings of Fernandez-Duque & Black [17] which assesses attention processes in Alzheimer's disease and in aging subjects. Their study uses a modified version of ANT which is varied to take into account the cost of disengaging from an invalid location. The modified version of ANT, in addition to a no-cue, cued and double (neutral) cue condition, also uses an invalid cue condition in which the cue appears in a location opposite to the target location.

The model was modified to incorporate the new invalid cue condition and, to reflect the changes in attention network functionality demonstrated in these studies, the following changes were made. Orienting effect was altered by tuning the buffer stuffing mechanism of ACT-R by increasing the *screen-x* values that determine what will be placed in the visual buffers using the command *(set-visloc-default :screen-x (within 20 180) :attended new)*. This corresponds to a slower orienting effect because the *screen-x* values range wider compared to where the target is placed on the screen and there is a higher probability of choosing a location other than the center arrow. The effect of lesioning the cognitive control network, which increases the congruency effect, is modelled by using productions that make the model refocus every time a distractor is picked up by mistake using production: *refocus-again-if-incongruent*. This results in an extra *move-attention* and thus the reaction time slows down. Similarly, in the case of an invalid cue, the model calls an extra production which shifts the focus of attention from the invalid location to the actual location of the target which takes more milliseconds compared to valid priming. The overall rule firing time (that is the ACT-R parameter *:dat*, the default activation time) is reset to 50 ms rather than 40 ms as used in the Wang et al model [53] and its reimplementation.

3.2   Results

The overall reaction times recorded by the model and compared with human data are given in Table 2. The model seems to fit the human data well with a correlation of 0.95.

Table 2. The reaction times for Alzheimer's disease(AD) subjects and model

	Congruent – AD Subject	Incongruent-AD Subject	Congurent-Model	Incongurent-model
Nocue	851	947	545	680
Uncued	817	982	545	680
Cued	729	889	488	599
Alert	761	958	520	614

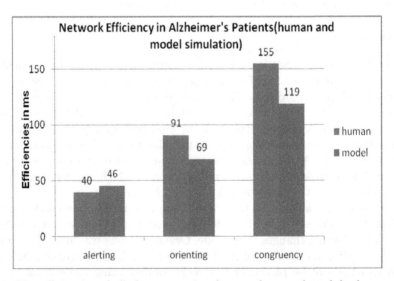

Fig. 3. The efficiencies of all three attentional networks are plotted for human data [17] vs the model simulation in ACT-R 6.0. The correlation of the efficiencies is 0.99.

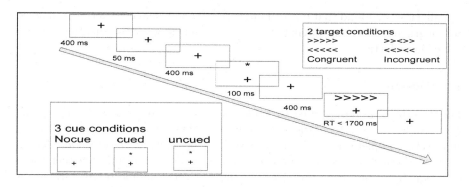

Fig. 4. A sketch of the design of the adapted version of ANT [10]

The efficiency of each network is measured using formula (4) from the original experiment and formulae (5)–(6). Figure 3 compares the model generated results with the human study results. As reported in the human study [17], the alerting cue increased the congruency effect but the presence of a spatially valid cue was ineffective in reducing the cost of incongruency.

$$Alerting\ efficiency = RT(no\text{-}cue) - RT\ (neutral) \tag{5}$$

$$Orienting\ efficiency = RT(uncued) - RT(cued) \tag{6}$$

The results reproduced by the model are in line with the findings of experiments studying attention related deficits in Alzheimer's patients. The model may potentially be used to see how the networks modulate each other and whether enhancing one network could make up for deficit in the other [9]. These results can be compared with the simulated results of the un-lesioned model, to demonstrate the inhibitory efffect of the attentional networks.

4   Modulation Effects of Alerting, Orienting and Executive Control

In the original ANT it is difficult to study the interactions of networks since the alerting and orienting effects have been measured using the same variable; that is, spatial cueing is used for orienting whereas temporal cueing is used for alerting. In order to clearly identify the modulating effect of one network on the other, Callejas and colleagues [9,10] modified the ANT using a separate tone for alerting whilst retaining cueing for orienting, as illustrated in Figure 4. An alerting sound was added to the original design of Fan et al. [16], and the new cueing variables used were: no-cue, where the stimulus is not preceded by a cue; cued, where a spatial cue is presented in the location where the stimulus is expected; and un-cued, where a cue appears in a location opposite to the location of the stimulus (invalid priming).

Callejas et al. found that both auditory signal and visual cue exert an influence on congruency; alerting having an inhibitory effect whereas orienting has an enhancing effect.

4.1 Design

The experimental design used to model this study involves 2 (auditory signal) x 3 (visual cue) x 2 (congruency) conditions. The symbolic component of the architecture implements each condition using rules such as *detect-sound, notice-stimulus-at-cued-top-location*, and so on. The model determines whether there is a high frequency tone produced by the auditory module of ACT-R and a flag is set indicating whether alerting is present or absent. Also, depending on the cue type, the model presents it on the screen; if its nocue, then the target appears without being preceded by a cue. In the case of the cued condition, the arrows are presented in the expected correct target location and, in the case of un-cued, the model presents the arrows in the incorrect location opposite to that of the expected target location. In the nocue condition, the model has an extra production which handles the 'surprise' condition where the target appears without any priming effect.

In the case of no alerting sound, the model implements an extra production which makes the system do an additional state change which increases the overall reaction time. In the case of an alerting signal, no such state switching is required. Similarly, in the uncued condition, an extra move-attention is required to move focus to the actual target location, whereas in the cued condition, the focus is already at the target location which saves milliseconds. The sub-symbolic component of ACT-R implements the attentional networks by using utility values and noise to help the model resolve conflicts and also make human-like errors. Incongruency is handled by two identical productions namely *refocus-again-if-incongruent* and *harvest-target-directly-if-incongruent* with different utility values (utility values are described in section 2.1).

4.2 Results

The overall reaction times recorded by the model compared with human data are given in Table 3. Pearson correlation coefficient was used to measure the degree of linear correlation between the two results. The coefficients came out to be 0.89 giving a good fit to the data.

The efficiency of each network is measured using formulae (4), (6) and (7).

$$Alerting\ efficiency = RT(no\text{-}alert) - RT\ (alerted) \tag{7}$$

The model showed similar interactions between the networks as in the original experiment in which the alerting network has an inhibitory influence on the congruency effect (cf. "clearing of consciousness" [37] p7401). Also, the orienting network had an influence on the control network; that is, when the location of the target was cued correctly, the congruency effect was smaller compared to the

Table 3. Results generated by the ACT-R model along with human data from Callejas et al. [9] in brackets

Mean Reaction Times for each condition for the experiment and (the model simulation)						
	No alerting tone			Alerting tone		
	No cue	Cued	Uncued	No cue	Cued	uncued
Congruent	573 (577)	533 (527)	561(595)	530 (545)	519 (475)	547 (545)
Incongruent	644 (690)	617 (597)	648 (710)	625 (680)	603 (543)	659 (680)

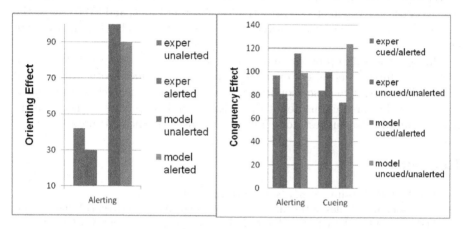

Fig. 5. Interactions between the variables. Congruency effect as a function of cueing and alerting; Orienting effect as a function of alerting.

condition in which the location of the target was cued in the opposite location. Interestingly, alerting speeded up the orienting of attention. The modulation effects of the attentional networks are illustrated in Figure 5.

These results can help us to understand not only how our attentional systems work but also explain how they function in a coordinated way to produce effective behavior. We are able to see how the control network can benefit from the work done by the orienting network in order to resolve conflict better and faster; the alerting system helps us prepare for a task and hence prevents the control network from doing processing work. Also, the orienting network can take advantage of this preparatory state of the system to speed up the orienting process. This clearly shows that, although these networks may be anatomically and functionally independent, they function under the influence of each other to produce effective behavior.

5   Conclusion and Future Work

The work described in this paper is based on the assertion that the whole attentional process comprises operations that help us to select a target found among distracters, to prepare ahead for an incoming stimulus so response is fast and correct, and to be able to resolve conflict and exert control whenever required. In the paper, through modeling the three components of attention, namely alerting, orienting and executive control, to jointly explain the cognitive phenomenon of attention, it seems we are approaching a more holistic view of the mechanisms of selective attention. The purpose of ACT-R models described in this paper is three-fold: (1) to facilitate simulating the behavioral study so that further predictions can be made; (2) to determine which networks may be affected or be functioning abnormally in attentional disorders in clinical patients, by simulating the effect of Alzheimer's on attention related conditions; and (3) to assess the behavior and efficiency of attentional networks and to study their modulation effects.

This work is still in progress and there are several areas that we would like to look into in further depth. For example we have plans to model other attentional related disorders, such as schizophrenia, in a similar fashion which may enable us to make further predictions about the behavior and efficiencies of the networks and potentially also suggest non-clinical methods of attention training.

References

1. The ACT-R Website, http://act-r.psy.cmu.edu/actr6/
2. Anderson, J.R.: Rules of the mind. Lawrence Erlbaum Associates, Hillsdale (1993)
3. Anderson, J.R., Bothell, D., Byrne, M.D., Douglass, S., Lebiere, C., Qin, Y.: An integrated theory of the mind. Psych. Rev. 111(4), 1036–1060 (2004)
4. Anderson, J.R., Lebeire, C.: The atomic components of thought. Lawrence Erlbaum, Mahwah (1998)
5. Anderson, J.R., Matessa, M., Douglass, S.: The ACT-R theory and visual attention. In: Proc. Seventeenth Ann. Conf. Cog. Sci. Soc., Nashville, Tennesse, pp. 61–65 (1995)
6. Booth, J., Carlson, C.L., Tucker, D.: Cognitive inattention in the ADHD subtypes. In: Proc. Tenth Mtg Int. Soc. Res. Child and Adolescent Psychopathology (ISR-CAP), Vancouver (2001)
7. Broadbent, D.E.: Perception and Communication. Pergamon Press, London (1958)
8. Bundesen, C.: A Theory of Visual Attention. Psych. Rev. 97, 523–547 (1990)
9. Callejas, A., Lupianez, J., Tudela, P.: The three attentional networks: on their independence and interactions. Brain Cognition 54, 225–227 (2004)
10. Callejas, A., Lupianez, J., Funes, M.J., Tudela, P.: Modulations among the alerting, orienting and executive control networks. Exp. Brain Res. 167, 27–37 (2005)
11. Cohen, J., McClelland, J.L., Dunbar, K.: Automatic processes in the Stroop effect: A parallel distributed processing account. Bull. Pscyhonomic Soc. 25, 239 (1987)
12. Corbetta, M., Shulman, G.L.: Control of goal-oriented and stimulus-driven attention in the brain. Nature Reviews 3(3), 201–215 (2002)
13. Droll, J.A., Gigone, K., Hayhoe, M.M.: Learning where to direct gaze during change detection. J. Vision 7(14), 1–12 (2007)
14. Eriksen, B.A., Eriksen, C.W.: Effects of noise letters upon the identification of a target letter in a non search task. Perception and Psychophysics 16, 143–149 (1974)
15. Fan, J., McCandliss, B.D., Fossella, J., Flombaum, J.I., Posner, M.I.: The activation of attentional networks. Neuroimaging 26, 471–479 (2005)

16. Fan, J., McCandliss, B.D., Sommer, T., Raz, M., Posner, M.I.: Testing the efficiency and independence of attentional networks. J. Cogn. Neurosci. 3(14), 340–347 (2002)
17. Fernandez-Duque, D., Black, S.E.: Attentional networks in normal aging and Alzheimer's disease. Neuropsychology 20(2), 133–143 (2006)
18. Gooding, D., Braun, J., Studer, J.: Attentional network task performance in patients with schizophrenia–spectrum disorders: Evidence of a specific deficit. Schizophrenia Res. 88(1-3), 169–178 (2006)
19. Gray, W.D., Young, R.M., Kirschenbaum, S.S.: Introduction to this special issue on cognitive architectures and human-computer interaction. Human-Computer Interaction 12, 301–309 (1997)
20. Heinke, D., Humphreys, G.W.: Computational Models of Visual Selective Attention: A Review. In: Houghton, G. (ed.) Connectionist Models in Psych., pp. 273–312. Psychology Press (2005)
21. Hopfinger, J.B., Buonocore, M.H., Mangun, G.R.: The neural mechanisms of top-down attentional control. Nat. Neurosc. 3, 284–291 (2000)
22. Howes, A., Young, R.M.: The role of cognitive architecture in modeling the user: Soar's learning mechanism. Human-Computer Interaction 12(4), 311–343 (1997)
23. Itti, L., Koch, C., Niebur, E.: A model of saliency-based visual attention for rapid scene analysis. IEEE Trans. Patt. Anal. Mach. Intell. 20, 1254–1259 (1998)
24. Itti, L., Koch, C.: Computational modeling of visual attention. Nature Reviews Neuroscience 2(3), 194–203 (2001)
25. Keiras, D.E., Meyer, D.E.: An overview of the EPIC architecture for cognition and performance with application to human-computer interaction. Human-computer Interaction 12, 391–438 (1997)
26. Klein, M.R.: Chronometric explorations of disordered minds. Trends Cog. Sci. 7(5), 190–191 (2003)
27. Koch, C., Ullman, S.: Shifts in selective visual attention: towards the underlying neural circuitry. Hum. Neurobiol. 4, 219–227 (1985)
28. Labarge, D.: Attention, awareness, and the triangular circuit. Consc. and Cog. 6, 140–181 (1997)
29. Laird, J.E., Newell, A., Rosenbloom, P.S.: Soar: An architecture for general intelligence. Art. Intell. 33(3), 1–64 (1987)
30. Lovett, M.C.: A Strategy-Based Interpretation of Stroop. Cog. Sci. 29(3), 493–524 (2005)
31. McClelland, J., Rumelhart, D.: An interactive activation model of context effects in letter perception: Part 1. An account of basic findings. Psych. Rev. 88, 375–407 (1981)
32. Navalpakkam, V., Arbib, M.A., Itti, L.: Attention and Scene Understanding. In: Itti, L., Rees, G., Tsotsos, J.K. (eds.) Neurobiology of Attention, pp. 197–203. Elsevier, San Diego (2005)
33. Newell, A.: Unified Theories of Cognition. Harvard University Press, Cambridge (1990)
34. O'Reilly, R.C., Munakata, Y.: Computational Explorations of Cognitive Neuroscience. MIT Press, Cambridge (2000)
35. Phaf, R.H., Van der Heidgen, A.H.C., Hudson, P.T.W.: SLAM: A connectionist model for attention in visual selection tasks. Cog. Psych. 22, 273–341 (1990)
36. Posner, M.I.: Orienting of Attention. Quart. J. Exp. Psych. 32, 3–25 (1980)
37. Posner, M.I.: Attention: the mechanism of consciousness. Proc. Nat. Acad. Sci. 91(16), 7398–7402 (1994)
38. Posner, M.I., Boies, S.J.: Components of attention. Psych. Rev. 78, 391–408 (1971)

39. Posner, M.I., Fan, J.: Attention as an organ system. In: Pomerantz, J. (ed.) Neurobiology of Perception and Communication: From Synapse to Society. De Lange Conference IV. Camb. Univ. Press, London (2007)
40. Posner, M.I., Petersen, S.E.: The attention system of the human brain. Ann. Rev. Neurosc. 13, 25–42 (1990)
41. Posner, M.I., Rothbart, M.K.: Research on attention networks as a model for the integration of psychological science. Ann. Rev. Psych. 58, 1–23 (2007)
42. Posner, M.I., Rothbart, M.K., Vizueta, N., Levy, K.N., Evans, D.E., Thomas, K.M., Clarkin, J.F.: Attentional mechanisms of borderline personality disorder. Proc. Nat. Acad. Sci. 99(25), 16366–16370 (2002)
43. Posner, M.I., Sheese, B.E., Odludas, Y., Tang, Y.: Analyzing and shaping human attentional networks. Neur. Networks 19, 1422–1429 (2006)
44. Raz, A.: Anatomy of Attentional Networks. Anatom. Rec., Part B, New Anatomist 281(1), 21–36 (2004)
45. Rensink, R.A.: The dynamic representation of scenes. Visual Cognition 7, 17–42 (2000)
46. The Sackler Institute for Developmental Psychobiology,
    http://www.sacklerinstitute.org/users/jin.fan/
47. Schneider, W., Shiffrin, R.M.: Controlled and automatic human information processing: I. Detection, search and attention. Psych. Rev. 84, 1–66 (1977)
48. Torralba, A.: Modeling global scene factors in attention. J. Opt. Soc. Am., Special Issue on Bayesian and Statistical Approaches to Vision 20(7), 1407–1418 (2003)
49. Torralba, A., Oliva, A., Castelhano, M., Henderson, J.M.: Contextual Guidance of Attention in Natural scenes: The role of Global features on object search. Psych. Rev. 113(4), 766–786 (2006)
50. Treisman, A.M., Gelade, G.: A feature-integration theory of attention. Cog. Psych. 12, 97–136 (1980)
51. Treisman, A.M., Sato, S.: Conjunction search revisited. J. Exp. Psych.: Hum. Percept. and Perf. 16, 459–478 (1990)
52. Tsotsos, J.K., Culhane, S., Wai, W., Lai, Y., Davis, N., Nuflo, F.: Modeling visual attention via selective tuning. Art. Intell. 78(1-2), 507–547 (1995)
53. Wang, H., Fan, J., Johnson, T.R.: A symbolic model of human attentional networks. Cog. Sys. Res. 5, 119–134 (2004)
54. Wang, H., Fan, J., Yang, Y.: Toward a Multilevel Analysis of Human Attentional Networks. In: Forbus, K., Genter, D., Regier, T. (eds.) Proc. 26th Ann. Conf. Cog. Sci. Soc., Chicago, Illinois, pp. 1428–1433 (2004)
55. Wang, K., Fan, J., Dong, Y., Wang, C., Lee, T.M.C., Posner, M.I.: Selective Impairment of attentional networks of orienting and executive control in Schizophrenia. Schizophrenia Res. 78, 235–241 (2005)
56. Wang, H., Fan, J.: Human attentional networks: A connectionist model. J. Cog. Neurosc. 16(10), 1678–1689 (2007)
57. Wolfe, J.M.: Guided Search 2.0: A revised model of visual search. Psychonomic Bull. and Rev. 1(2), 202–238 (1994)

The JAMF Attention Modelling Framework

Johannes Steger, Niklas Wilming, Felix Wolfsteller,
Nicolas Höning, and Peter König

Neurobiopsychology Group,
Institute of Cognitive Science, University of Osnabrück
{jsteger,nwilming,fwolfste,nhoening,pkoenig}@uos.de

Abstract. Many models of attention have been implemented in recent years, but comparison and further development are difficult due to the lack of a common platform. We present JAMF, an open source simulation framework for drag & drop design and high-performance execution of attention models. Its building blocks are "Components", functional units encapsulating specific algorithms. Simulations are created in the graphical *JAMF client* by connecting Components from the server's repository. Today it contains Components suitable for replication and extension of many major models of attention. Simulations are executed on the *JAMF server* by translation of model definitions into binary applications, while automatically exploiting the model's structure for parallel execution. By disentangling design and algorithmic implementation, the JAMF architecture combines a novel tool for rapid test and implementation of attention models with a high-performance execution engine.

Keywords: Attention, Modelling, Saliency, Simulation, Software.

1 Introduction

In recent years many models of overt attention have been proposed, employing a variety of methods, such as Bayesian techniques [1], neural networks [2], machine learning algorithms [3] and saliency approaches [4].

To evaluate the performance of such models, their predictions must be compared to real-world data. In the visual domain these data are generally acquired by means of eye-tracking, with measurement of many subjects' viewing behaviour used to yield a stimulus-specific signature of overt attention. In order to generate data from attention models, they first have to be instantiated as computer simulations.

The latter in particular is a non-trivial task lying at the interface of computer and neuroscience: Modelling attentional processes necessitates deep understanding of the subject matter, while efficient algorithmic implementation requires advanced computer science skills. This broad set of demands makes the task difficult for an expert from either of these domains.

A practical solution is to use existing attention simulations, such as the iLab Neuromorphic Vision C++ Toolkit [5], as a starting point. This can indeed drastically reduce the amount of code which has to be implemented,

L. Paletta and J.K. Tsotsos (Eds.): WAPCV 2008, LNAI 5395, pp. 153–165, 2009.
© Springer-Verlag Berlin Heidelberg 2009

but introduces several other problems. First of all, learning how to use a new framework can involve a steep learning curve, especially for non-computer scientists, and may require the user to read a large amount of source code. Second, using an existing source-code base may even introduce new complexity. The code-base might be optimal for a specific modelling approach, but its structure might impose limitations on the type of models that can be implemented. Third, not all attention model implementations are freely available, and such algorithms must be re-implemented by the user. The question still remains then, of how easy it is to combine, alter and adapt existing models to one's own needs.

In this paper we present JAMF, a generic attention modelling framework built to address these issues. It is free software and released under the GNU Public License (version 2). At the heart of the framework lies a separation in the implementation of the attention model itself and the supporting technical aspects. Models are composed of functional units, called *Components*, arranged as a directed graph. Such model graphs abstract from the programming code involved in implementation, and allow for purely graphical creation of models. When using JAMF, the designer does not need to be concerned with any implementation details. Additionally, Components exist independently of the models in which they are used, allowing for efficient reuse of existing code. The encapsulation of Components allows the framework to automatically and transparently optimize simulations for use in multi-processor environments and leverage available performance libraries. The existing JAMF Component base provides a useful selection of the standard functionality needed in many attention models, e.g. feature extraction and machine learning algorithms. Development, code review and release of Components is coordinated via a version control system, bug tracker, and mailing list. All of these sources can be accessed on the project's website, which provides additional documentation and support.[1]

The framework was designed with the express aim of integration into existing working environments, and provides import and export functionality for standard mathematical analysis tools on the model developer side. Existing algorithms can easily be turned into new Components – if they are already implemented in C(++), a lightweight wrapper interface is all that is needed. If they are available as Matlab functions, they can be wrapped in a special Component for direct inclusion. Overall, we believe JAMF is an accessible and useful framework for building working models of attention.

In the next section we describe the technical aspects of the JAMF attention modelling framework, and it should be noted that such details are needed only by Component developers. Afterwards we provide three case studies involving different user scenarios, demonstrating how JAMF can be used by neuroscientists. We close with a comparison of JAMF to existing modelling frameworks.

---

[1] http://jamf.eu/

2   Design of JAMF

2.1   System Overview

In response to the requirements of easy-to-use model design and fast execution of the resulting simulation, JAMF was designed as a client/server architecture (see Figure 2 for an overview). The server part is written in C/C++ and contains the functionality for running a simulation, as well as algorithmic building blocks in the form of Components. The client part is written in Python and used for design, parameterization, control of simulations, and data import and export. It is designed to act as a front-end to the server, transparently encapsulating all of its functionality. A screen shot of the client's graphical interface is displayed in Figure 1. The client can also be accessed directly, for example via an interactive Python shell or from user-written programs.

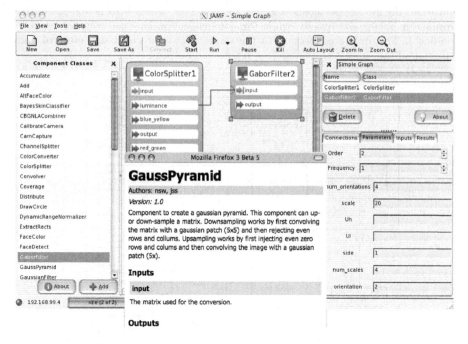

Fig. 1. The JAMF client user interface, with the main window in the background. The "Component Classes" panel on the left lists all Components available on the server. From here, Components are instantiated and added to the graph canvas (centre panel), where their input and output methods can be graphically connected. The right panel displays properties of the currently selected Component and allows for specification of parameters and inputs, as well as introspection of the Component's output. The top panel contains buttons allowing local data management and remote control of a simulation. The foreground window shows the HTML documentation for a Component. It is directly accessible through the "About" button below the Component list in the main window.

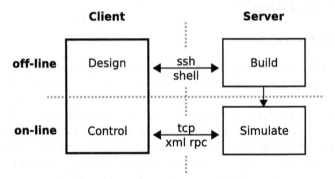

Fig. 2. JAMF architecture overview. By mirroring the server's Component interface, the client acts as a transparent proxy to all server functions. The server side on the right provides the Component repository. On startup, its interface description is read by the client over SSH ("off-line") and used for **design** of new models. During execution of a simulation, the model specification is sent to the server, which **builds** and starts the **simulation** binary accordingly, while the communication channel switches to TCP ("on-line"). This online connection to the server allows for client-side **control** and inspection of the simulation parameters, inputs and outputs.

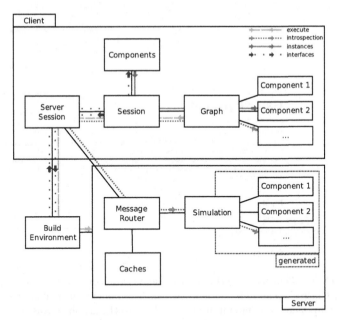

Fig. 3. Simplified communication diagram: Boxes represent (aggregated) system Components, with their associations depicted by black lines. The paths outline information flow: request and retrieval of Component interface descriptions (interfaces); instantiation of Component classes for graph design (instances); source code generation and simulation execution (execution); remote control and introspection of simulation (introspection). The dotted box shows the portion of the server generated dynamically on a per-graph basis.

Client and server can run on the same physical machine, or may alternatively be distributed on a network. As well as minimising installation and maintenance expenses, the distributed approach allows for the use of a (possibly remote) dedicated computing machine to execute the server simulations, while users can work locally using the graphical interface. Communication between server and client works over two channels: SSH and TCP/IP. SSH is used in the pre-simulation environment, and TCP/IP becomes available when the simulation is executed.

Next we will outline the basic design of the framework and highlight some of its main features. The structural relations of the architecture's constituents are depicted in Figure 3. Colored lines show information flow, and in the following sections we will first describe the C implementation of Components, describe how their interface description is read by the client (blue) and used as the basis for building new model graphs (red). We will then show how those are translated into simulations on the server side, executed (purple) and remotely controlled by the client (green).

2.2   Components

Server Implementation. Components are designed as C++ wrappers around C algorithms (see Figure 4 for a source code sample). All communication between them is carried out exclusively by passing pointers to instances of OpenCV's Matrix type "CvMat", a flexible structure that can represent matrices using any basic numeric type.[2]

This results in a very simple interface, where each Component defines one or more input methods that take exactly one matrix and store it in a private variable, while output methods return one matrix each. Given the power of the matrix type, the simplicity of the method's signature does not impose any limits on inter-Component communication. To expose parameters, a Component can implement setters that store values similarly to input methods described above. Parameter types are restricted to **bool**, **int**, **double** and **string**.

Note that input, output and parameter methods provide data access only – the actual algorithms are implemented in a Component's **run()** method. This method starts with a set of preconditions, which are assertions about input formats and parameter values. Memory management within one Component is dealt with exclusively in the **run()** method and is limited to the output data. This method is also responsible for releasing any memory that is still allocated from previous invocations.

This construction guarantees maximal encapsulation of algorithms while still allowing for robust automated optimization. The base Component "BasicComponent" defines default input and output functions, hence only implementation of the **run()** method is necessary to create a simple Component (see Figure 4).

---

[2] OpenCV (http://opencvlibrary.sourceforge.net/) is a popular Open Computer Vision library originally developed by Intel. It supports the Intel Performance Primitives (IPP) for architecture-specific optimizations.

```
1   class TimesTwo : BasicComponent
2   {
3   public:
4       run() {
5           PRE_INPUT(in);
6           // create output matrix
7           if (out)
8               cvReleaseMat(&out);
9           out = cvCreateMat(in->type, in->rows, in->cols);
10          // do the calculation
11          cvScale(in, out, 2);
12      }
13  }
```

Fig. 4. Minimal Component source code: Using the default input and output channel inherited from BasicComponent, only the algorithm logic is left to be implemented in the Component's run() method

To extract interface information, JAMF runs gccxml[3] on Components' header files and filters for valid method signatures. In addition to this interface description, each Component is accompanied by an XML file providing end-user documentation that is rendered to HTML (PNG for formulae) for easy viewing. The interface description and help files of all server Components are bundled into a single "components.zip" file that acts as a complete description of the server's Component repository.

Client. In order to use the server's Components for model design, the client must access the Component interface description generated on the server. When establishing an SSH connection to a server, the client automatically updates its local Component description file – for every Component on the server side, it dynamically generates a Python class mirroring the Component's interface (blue message line in Figure 3). It has setters parallel to a Component's inputs and parameters that are type aware, and output methods that can be queried for results. In addition, each Component is accompanied by HTML documentation describing its general function and providing details on inputs, outputs and parameters (c.f. 1). All in all, the Client wraps a server Component in a convenient Python interface and sets up hooks to propagate changes directly to a running server (see Section 2.5). Adding a Component to a model is easily done, simply by instantiating such a Python class (red messages in Figure 3).

2.3 Model Graph

Output methods of instantiated Components can be connected to input methods, thereby forming a directed model graph. When executed, the server will feed data along these connections, passing each Component once, making up one full

---

[3] gccxml (http://gccxml.org/) dumps gcc's internal representation of a source file as an XML description.

"iteration". Some Components yield a dynamic number of outputs or accept a dynamic number of inputs. Subgraphs between such Components are executed once for each output matrix. By default, the client will determine the execution order by traversing the graph starting from all Components without connected inputs and assigning an increasing "order". Where parallel execution is possible, for example when two Components do not depend on each others' outputs, the same "order" will be assigned. In addition, Components can request to be called with a certain "frequency" with respect to the server iteration and an "offset" to the start iteration. A set of parameterized and connected Components makes up a model graph.

2.4   Build

To allow for later execution of the model, the source code for the simulation first must be generated and compiled. The client's representation of the model is translated to an XML specification of the graph structure (purple line in Figure 3). This XML file is sent to the server and translated into C code by XSLT[4]. In this translation process, the Component's "order" property is used to create OpenMP[5] parallel regions where possible, thereby providing transparent multi-processor optimization (see Figure 5 for a sample graph). The generated code is embedded directly in the server's simulation class and linked against the Component repository library. After successful code generation, the client can build the simulation binary corresponding to its model graph.

2.5   Execution

After building the server binary for a specific attention model, the client can start a simulation over the existing SSH channel. Upon startup, the server sets up threads for TCP listening, message routing and the simulation. The client can then login via TCP and send remote procedure calls (RPC) encoded in XML. On the server side, XML RPC messages are converted into C structs and routed to their target thread, as shown by the green line in Figure 3. Messages are used for a number of purposes: to address the simulation directly for the control of model execution; to deliver parameters and inputs to Components; or to request intro-spection data by querying Components' output methods, with answers routed back to the client. On the client side, these messages are generated by calling the respective methods of the Component instance. Parameters and inputs are set and sent for current or future iterations, allowing the server to be programmed in advance. The server also accepts input filenames in the form of URLs using the HTTP, FTP, NFS or Samba protocols. Whenever such a filename is detected, the server's "incache" will grab the file from the source location, save it in a

---

[4] XSLT is an XML based language for transformation of XML documents into other text formats.

[5] OpenMP (http://www.openmp.org/) is a multi-platform shared-memory parallel programming API supported by many major compilers, including GNU's gcc 4.2 and Intel's icc.

local cache and optionally send a "start" command to the simulation on success. For the outputs delivered by the server, the client offers immediate inspection of graphical data as well as export to a variety of formats (txt, matlab, bitmap and vector image formats, pdf). To facilitate client-independent operation, the server optionally offers a cache for all outgoing messages ("outcache") by serializing them to disk, which allows clients to retrieve data from past iterations.

The core client engine used by the user interface can also be employed by the user to write powerful scripts, allowing the engine to be used, for example, to harvest a more complex Component parameter space or to run a batch job over structured stimulus material.

3   Case Studies

In order to show the broad applicability of the JAMF framework, we here introduce three user case studies. First, we show how a probabilistic attention model can be implemented. Second, we outline how JAMF can be used to control a robot head equipped with stereo vision. And third, we elaborate on how useful it can be for students learning about different models of attention.

3.1   Bayesian Saliency Model

Eye tracking data shows that local features contribute to visual attention in specific manners. In contrast to luminance, color contrasts for instance do not contribute linearly to salience, rather exceptionally high and low color contrasts are usually far more informative in predicting salience than intermediate level. A recent Bayesian saliency model uses a large amount of eye-tracking data to empirically capture the specific relation in which salience depends on the feature values without making parametric assumptions [1].

The Bayesian model describes the salience of a feature as the conditional probability to fixate an image location given the feature value at the location. Salience as conditional fixation probability is measured in the ratio between the likelihood with which feature values occur at empirically fixated points and the respective prior probability with which they occur in the image. In other words, salience of a particular feature value is defined in the proportion to which this value of the feature occurred more often at fixations than in the image. This particular model further assumes a whitening of features values according to the distribution of features in natural images, as has been shown for contrasts in the LGN. Hence before computing conditional fixation probabilities, absolute feature values of the stimuli are 'whitened' by rank-ordering in 20 bins with equal amounts of values (considering all feature values in the image database of the study).

We implemented this attention model in a JAMF simulation. The model graph for the Bayesian salience model is depicted in Figure 5. All Components necessary for the simulation are contained within the JAMF standard Component library. To analyze color-features, images have to be coded in independent intensity and color channels, such as in the HSV color space. Here we use the

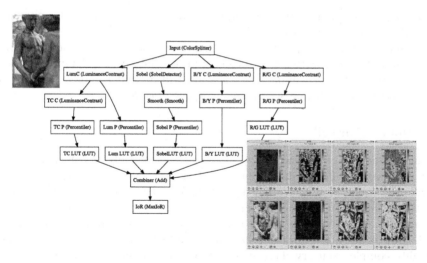

Fig. 5. Conditional feature probability graph. The first three layers in the graph shown here deal with the extraction of image features. Percentiles are computed in the fourth layer, which are matched to feature probabilities in the fifth layer. Components within each layer are executed in parallel. The last two layers combine probabilities and normalize the result. The inset figure (bottom right) shows intermediate results. Top row, from left to right: Red/Green contrast, 2nd order contrast, percentiles and feature probabilities. Second row: Image luminance, Sobel edge detector, percentiles and feature probabilities.

ColorSplitter Component to transform bitmap images into the the Derrington-Krauskopf-Lennie (DKL) color space, which is a physiological color space defined by relative excitations of the three cone types in the retina (L, M, and S). In brief it yields channels for luminance intensity, "constant blue" (or "yellow-blue") and "tritanopic confusion" (or "red-green"). By drag and drop, the respective intensity and color channels are connected to various Components to obtain a master saliency map. *LuminanceContrast* and *Sobel* Components calculate luminance, color and texture contrast as well as edge intensities. The output of these Components is again connected to the *Percentiler* Component, which performs the whitening of the stimuli statistics for the respective feature, which in turn is fed into the *LUT* Component to compute empirical within-channel saliency maps using previously measured conditional fixation probabilities. Finally, these maps are fed into the *Combiner* Component to obtain a master saliency map by additive integration. Note that only this last Component has to be exchanged to implement the multiplicative integration scheme explored in the study of Schumann et al. [1].

There are several possible approaches to generate fixation sequences from the master saliency map. For example, Itti's [4] method of Winner-takes-all and Inhibition-of-return is available in the Review repository, together with a Component for non-linear amplification and combination of conspicuity maps. Both generative schemes were combined simply by adding the respective Component (*MaxIoR*) in the final processing stage.

In summary, this example shows how JAMF can be used as a tool for rapid implementation of innovative new modelling ideas. The existing Component base allows neuroscientists to implement their own ideas and combine them with existing approaches without writing a single line of code.

3.2   Robot Control

Attention models can be used to help robots interact sensibly and efficiently with their environment. For example, implementing a model of overt attention can provide the robot with active sensing capabilities. By identifying and tracking interesting targets in a scene, the model can also enhance scene understanding by directing sensors towards salient regions that can then be processed in higher detail. Covert attention, on the other hand, can help with real-time processing of highly complex sensory data. Most robots are equipped with only limited processing capabilities, so it is an advantage to be able to select those parts of the sensory input that are likely to yield the highest information gain.

JAMF's modular structure makes it possible to include interfaces to robot hardware as Components, both on the sensory side (IR sensors, microphones, laser scanners, etc.) and on the actuator side. Recently, JAMF was used to implement an attention-based controller on a stereo-vision robot head with 4 degrees of freedom. A rough sketch of the setup is shown in Figure 6. We developed a Component that grabs frames from the head's two cameras, and an actuator Component that transduces spatial target points to motor movements that tilt,

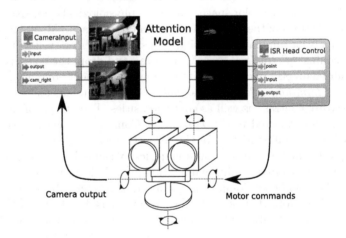

Fig. 6. Schematic depiction of how JAMF can be used to create an interface for a stereo-vision robot head. One Component retrieves video frames from the head's cameras, while another moves the head to target fixation points. The attention model that lies between these two Components generates target fixations, and is treated as a black box that can easily be exchanged.

pan and turn the head. The attention model between these Components analyses incoming camera images and generates new fixation targets for the head actuator Component. The implementation allowed the robot head to successfully track interesting objects defined by their "redness". Importantly, in such robotic setups the attention model can be treated as a black box and is easily interchangeable with more complex models. For example, the existing disparity Component could be "plugged in" to provide depth features.

The Robot Control use-case demonstrates that it is straight-forward to transfer attention models developed in JAMF to real-world platforms.

3.3 Classroom

Students studying existing attention models usually do so at a purely theoretical level. There is however a qualitative difference between knowing what a model does and seeing what it does. JAMF can help to bridge this gap between theory and practice. The graphical client allows students to interact with an attention model, easily explore the effects of different parameters and see how the model reacts to different kinds of inputs. A live view of intermediate results at different processing stages also makes it easier to understand what is happening at each step. JAMF gives students the opportunity to develop and understand models without requiring prior programming skills.

4 Related Work

JAMF is not the only software in the field of attention resarch – many researchers publish their code. Most often, however, only prototypes or partial implementations in high-level languages such as Matlab are provided. Among the fully fledged frameworks that are available, we picked four packages which solve similar problems or solve problems in a similar way. We give a short description of each and point out differences in a feature matrix (Table 1).

The *iLab Neuromorphic Vision C++ Toolkit* is developed at the University of California and at Caltech.[6] Over the years, many saliency models and extensions such as Bayesian surprise [6] have been implemented in the toolkit, by the original authors as well as by third party contributors. It comprises a set of C++ classes implementing a range of vision algorithms for use in attention models. However, as there is no higher-level design interface and the toolkit architecture is quite complex, advanced programming skills are required to leverage the toolkit's features. Notably, it also offers distributed execution support for Beowulf clusters. Although it is essentially inaccessible for non-programmers when compared to JAMF, the number of available algorithms and their optimizations make the iLab Toolkit an interesting option for the computer vision scientist.

LabVIEW [7] is a commercial software developed by National Instruments. It is based on G, a graphical dataflow programming language. Similarly to JAMF, a compiler is used to produce machine code from a graphical model. Furthermore,

---

[6] http://ilab.usc.edu/

Table 1. A feature matrix of comparable applications. "Visual" indicates whether graphical model development is supported, "Language" lists programming languages supported for extending simulations, "Remote" indicates if the simulation can be controlled and executed remotely. The "Opt." and "Parallel" columns list employed optimization and parallelization techniques.

Name	License	Visual	Language	Remote	Opt.	Parallel
iLab	GPL	-	C(++)	-		Beowulf, manual
JAMF	GPL	+	C(++), M	+	IPP	OpenMP, fully automatic
LabVIEW NI	commercial	+	G	-		proprietary
MatLab/ SimuLink	commercial	-/+	M, C, Fortran, Ada	-	IPP	proprietary
TarzaNN	"Open Source"	+	C(++)	-		Cluster, manual

the additonal Machine Vision Module comes with a wide range of functionality that suffices for standard machine vision tasks, but does not include any of the recently developed methods. Besides licensing costs and the proprietary nature of the G language, understanding its non-standard language paradigm also places an additional burden on new users.

Matlab is a commercial multi-purpose numerical computing environment developed by The MathWorks. It is based on the matrix-centered M language and in broad use in education and science. Key advantages are the wide variety of implemented algorithms and the superb documentation. Toolboxes that extend Matlab to solve more specific problems are available seperately. One of them is the commercial *SimuLink* package, an advanced simulation environment for dynamic systems, which allows graphical modelling. Although it could be used for graphical attention modelling, there are no appropriate Components supplied for this purpose. The disadvantages associated with Matlab and SimuLink are their high price and the learning curve associated with programming in M.

TarzaNN [8] was initially developed as a joint effort by Albert Rothenstein and Andrei Zaharescu at York University in Toronto for implementation of the selective tuning model [2]. It is a neural network simulator that abstracts from single neurons to layers of neurons, and was designed specifically to implement visual attention models. Extension of the simulator is limited to the implementation of filter functions for neurons. In contrast to JAMF, TarzanNN is only suitable for modelling a specific subset of today's attention models.

5   Conclusion

JAMF offers a unique combination of highly-optimized algorithmic infrastructure and an easy-to-use graphical modelling interface. The existing repository of Components is sufficient for building a wide range of model types, and while it allows the non-programmer to rapidly implement and run simulations of visual attention, it is also easily extended with new Components. Here we have demonstrated that the framework can be used as a suitable tool for such diverse tasks as the implementation of a novel attention model, or the attentional control of a stereo-vision robot head. Future work will include further extensions of the Component base, for example with algorithms for processing auditory information. Furthermore, we plan to use JAMF to support the development of a multi-modal attention model for the stereo-vision robot head.

Acknowledgements

This work is partially funded by the European Commission under the Cognitive Systems project POP (FP6-IST-2004-027268). We thank Prof. Helder Araújo João Xavier and Rui Caseiro from the Institute of Systems and Robotics in Coimbra, Portugal for their support and the chance to test JAMF on a stereo vision robot head.

References

1. Schumann, F., Acik, A., Onat, S., König, P.: Integration of different features in guiding eye-movements. In: Proceedings of the 7th Meeting of the German Neuroscience Society / 31th Göttingen Neurobiology Conference, Neuroforum 2007, Göttingen, Germany (2007)
2. Tsotsos, J., Culhane, S., Kei Wai, W., Lai, Y., Davis, N., Nuflo, F.: Modeling visual attention via selective tuning. Artificial Intelligence 78(1-2), 507–545 (1995)
3. Kienzle, W., Wichmann, F., Scholkopf, B., Franz, M.: Learning an Interest Operator from Human Eye Movements. In: Proceedings of the 2006 Conference on Computer Vision and Pattern Recognition Workshop (2006)
4. Itti, L., Koch, C., Niebur, E., et al.: A model of saliency-based visual attention for rapid scene analysis. IEEE Transactions on Pattern Analysis and Machine Intelligence 20(11), 1254–1259 (1998)
5. Itti, L.: The ilab neuromorphic vision C++ toolkit: Free tools for the next generation of vision algorithms (2004)
6. Itti, L., Baldi, P.: A principled approach to detecting surprising events in video. In: Proc. IEEE Conference on Computer Vision and Pattern Recognition (CVPR), San Diego, CA, pp. 631–637 (June 2005)
7. Johnson, G., Jennings, R.: LabVIEW Graphical Programming. McGraw-Hill Professional, New York (2001)
8. Rothenstein, A., Zaharescu, A., Tsotsos, J.: TarzaNN: A general purpose neural network simulator for visual attention modeling. In: Paletta, L., Tsotsos, J.K., Rome, E., Humphreys, G.W. (eds.) WAPCV 2004. LNCS, vol. 3368, pp. 159–167. Springer, Heidelberg (2005)

Modeling Attention and Perceptual Grouping to Salient Objects

Thomas Geerinck[1], Hichem Sahli[1], David Henderickx[2],
Iris Vanhamel[1], and Valentin Enescu[1]

[1] Electronics & Informatics Department (ETRO)
Interdisciplinary Institute for BroadBand Technology (IBBT)
[2] Faculty of Psychology and Educational Sciences (PE)
Vrije Universiteit Brussel (VUB)
Pleinlaan, 2 B-1050 Brussels, Belgium
{tgeerinc,hsahli}@etro.vub.ac.be
http://www.etro.vub.ac.be

Abstract. In this paper, we propose a biologically inspired model of
the middle stages of attention, with specific algorithmic details. Existing computational models of attention concentrate on their role in visual
feature extraction and the selection of spatial regions. However, these
methods ignore the role of attention in other stages. Extension of these
models has been proposed by augmenting the unit of attentional selection
to "proto-objects". In our approach, we extend attention to the middle
stages and integrate the selection process with the perceptual grouping
process. Integration is achieved by our innovative saliency driven perceptual grouping strategy, extending the traditional pixel-based saliency
map to salient proto-objects. The proposed selective attention is made
in two stages. Firstly, to achieve salient region localization, our method
enhances the saliency map with region information from image segmentation and selects the most salient region (proto-object). Then, regions
are organized using perceptual groupings, and their pop-out sequence is
determined. Compared with traditional attention models our model provides saliency maps with meaningful region information, by eliminating
misleading high-contrast edges, and focus of attention shifts in unit of
perceptual object rather than spatial region. These two improvements fit
to high stage vision information processing such as object recognition.
Experiments in a reduced set of images show that our proposed model
is able to automatically detect meaningful proto-objects.

1 Introduction

It is well known that the human visual system employs an attention mechanism,
due to limited processing resources, to selectively process important information
that is currently relevant to visual behaviors or visual tasks [1,2,3]. This mechanism deals efficiently with the balance between computing resources, time cost
and fulfilling different visual tasks in normal, cluttered or dynamic environments.

L. Paletta and J.K. Tsotsos (Eds.): WAPCV 2008, LNAI 5395, pp. 166–182, 2009.
© Springer-Verlag Berlin Heidelberg 2009

Selective visual attention provides the brain with a mechanism of focusing computational resources on one object at a time, either driven by low-level image properties (bottom-up attention) or based on a specific task (top-down attention). Moving the focus of attention to locations one by one enables sequential recognition of objects at these locations. What may appear to be a straightforward sequence of processes (first focus attention to a location, then process object information) is in fact an intricate system of interactions between visual attention and object recognition [3].

To date, there have been a number of attentional models for psychophysics or for machine vision that use the hypothesis of the "spotlight" or "zoom-lens" analogy for visual attention. Most of them are derived from Treisman's Feature Integrated theory [4]. However, traditional models have only concentrated on mechanisms of visual attention based on selectivity by spatial locations. They inherently lack mechanism to account for object-based visual selection, and hence are not perfectly suited to work in real-world natural scenes.

Three different requirements of attention have been identified [5]:

1. attention may need to work in discontinuous spatial regions or locations at the same time
2. attention may need to select an object composed of different visual features but from the same region of space
3. attention may need to select objects, locations, and/or visual features as well as their groupings for some structured objects.

For applying attention mechanisms in real and normal scenes, a computational approach inspired by the alternative theory of object-based attention is necessary [6]. In contrast to the traditional theory of space-based attention, object-based attention suggests that visual attention can directly select discrete objects rather than only and always continuous spatial locations within the visual field.

Recently, there has been a rapidly increasing interest in object-based attention (c.f. [3,7,5,8,9,10]) but research into useful systematic theories is still a very open area, especially practical models of object-based attention for real-world applications. For example in [8] a method is proposed for salient spatial area (as opposed to object) determination, used to guide object recognition. Several important issues must be addressed clearly in this context:

- Early identification and segmentation of perceptual objects
- The relationship between object-based and space-based attention
- Grouping/segmentation and object-based attention. A grouping is a hierarchical structure of objects and space, which is also the common concept in the literature of perceptual grouping. A grouping may be a point, an object, a region, or a structured grouping.
- Visual saliency and visual attention. The salience of a grouping measures how different this grouping contrasts with its surroundings and depends on various factors, such as feature properties, perceptual grouping, dissimilarity between the target and its neighborhood.

In this paper, following the above criteria, we propose a novel region-based focus of attention mechanism that is based on human attention and perceptual region grouping. As object-based image segmentation is beyond current computer vision techniques, the proposed method segments an image into regions, which are then merged using an innovative saliency-driven perceptual organization approach. At the same time, an attention region (AR) is created based on the saliency map and saliency regions from an image. A hierarchical perceptual grouping is used to select the salient regions, which are then clustered into the Object Of Interest (OOI) using a new region merging criteria. Unlike other algorithms, the proposed method allows multiple OOIs to be segmented according to the saliency map.

The contribution of this paper can be summarized in the following 3 aspects. Firstly, region is chosen as the perceptive unit, which makes the method more effective in terms of perception. Secondly, compared with traditional attention models our model provides saliency maps with meaningful region information, by eliminating misleading high-contrast edges. Finally using both global effect and contextual difference the proposed focus of attention shifts in unit of perceptual objects rather than spatial regions.

The remainder of the paper is organized as follows. Section 2 gives an overview of the proposed visual attention model. Sections 3 and 4 summarize the saliency map generation and the perceptual grouping, respectively. In section 5, we describe the proposed Inhibition of Return principle. Section 6 presents early results. Finally, section 7 gives some conclusions and future work.

2   Visual Attention Model

Figure 1 depicts the proposed the bottom-up model for visual attention. While sharing same concepts with existing models of visual attention, there are a number of differences concerning implementation details as well as structural design components that yielded the proposed object-based attention model.

On the input image, different *feature dimensions* (in the following simply called features) are computed: intensity, color, color opponency and orientation, in conformity with [10]. For each dimension, the pixel-based responses are computed on different scales and for different *feature types*, e.g. red, green, blue and yellow for the feature color. For each feature, we first compute an *image pyramid*, from which we compute *scale maps*. These represent saliencies on different scales for different feature types. The scale maps are fused into *feature maps* representing different feature types and these again are combined to *conspicuity maps*, one for each feature, thereby strengthening important aspects and ignoring others.

First, a hybrid approach combining low-level saliency and region information is used to produce enhanced conspicuity maps (enhanced C maps). The C maps are enhanced with region information (from image segmentation) by averaging the conspicuity values in each region. This approach has been also followed by [11] to produce region enhanced saliency. The output of this phase is a "multi-spectral" image combining all the enhanced conspicuity maps.

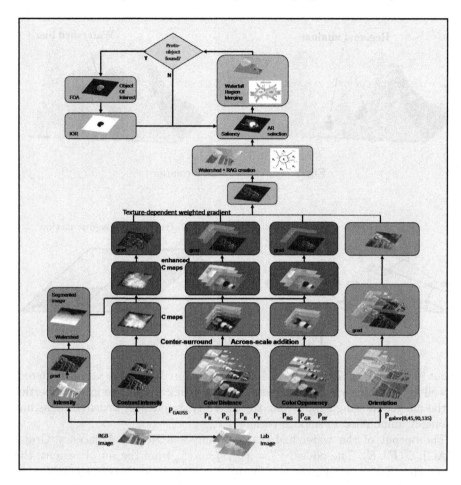

Fig. 1. Framework for modeling attention and perceptual grouping to salient objects of interest. (Waterfall illustration from [16])

Second, the obtained "multi-spectral" image is then segmented using the watershed transform. The watershed transform [12] is a morphological segmentation tool often applied on the gradient magnitude of an image in order to guide the watershed lines to follow the crest lines and the real boundaries of the regions. The idea, in its simplest way, can be visualized by considering successive thresholds of a function producing horizontal cross sections of a relief. The watersheds will be the dividing lines of the 'touching' cross sections. Figure 2 provides an illustration. The gradient of the "multi-spectral" image is obtained by combining, using the approach of [13], the gradients of the texture (from the orientation responses) and the gradients of the enhanced conspicuity maps. This approach allows obtaining a final gradient capturing all perceptual edges in the input RGB image. The method is general, in the sense that it makes limited assumptions

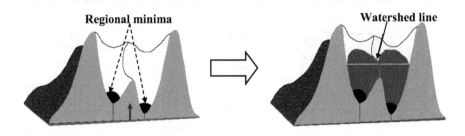

Fig. 2. Watershed transformation

Fig. 3. Region Adjacency Graph

about scene content. Both textured and non-textured areas are accommodated, as well as the region size is irrelevant. The processing adapted to local properties of the image. allowing suppressing the intensity gradient in textured areas but leaving it unmodified in smooth regions.

The output of the watershed segmentation is a Region Adjacency Graph (RAG), $G(P^0, E)$. The nodes, $P^0 = \{r_1^0, r_2^0, ..., r_{m_0}^0\}$ are the set of regions; the set of arcs E, connecting the nodes, are the boundaries between neighboring regions. Each region is represented by its average feature, which in this paper is the *Lab* color, $(\mu_L(r_i^0), \mu_a(r_i^0), \mu_b(r_i^0))$. A mosaic image (each region represented by its mean color) is also associated to the RAG. Figure 3 illustrates the RAG associated to a partition. Both are inputs to a region-based saliency map (section 3). The purpose of the saliency map is to represent the conspicuity of each location of the visual field, that is, salient regions extracted have higher prominent importance than the other regions. Subsequently, the region with highest saliency value is selected as attention region (AR) from the region saliency map.

Salient region extraction based on saliency map provides a good starting point for semantic-sensitive content representation. However, perceived salient region extraction for images is still an unsolved problem. One reason is that low-level features are often not enough to classify some regions unambiguously without the incorporation of high-level and human perceptual information into the classification process. Another reason for the problems is perception subjectivity. Different people can differ in their perception of high-level concepts, thus a closely related problem is that the uncertainty or ambiguity of classification in some regions cannot be resolved completely based on measurements methods.

In this paper, we propose a new method for prominent region extraction in images in order to remove the above mentioned limitations. The segmented image is analyzed by a number of perceptual attributes based on the *mise-en-scene* principles [14]. As a set of techniques, mise-en-scene helps compose the film shot in space and time, which are used by the film-makers to guide our attention across the screen, shaping our sense of the space that is represented and emphasizing certain parts of it. The used perceptual attributes are the contrast of a region with its surroundings, its Orientation Conspicuity, its compactness, and its *Compositional Balance* [14] which can be interpreted as the extent to which the areas of image space have equally distributed masses and points of interest.

Starting from the selected salient attention region (AR), an innovative saliency driven perceptual grouping of segmented regions is proposed to obtain perceptually meaningful regions. A meaningful image segmentation groups the pixels into disjoint regions that consist of uniform components. Facing absence of contextual knowledge, the only alternative which can enrich our knowledge concerning the significance of our segmented groups is the creation of a hierarchy guided by the knowledge which emerges from the superficial and deep image structure.

Our main goal here, is to create a hierarchy among the gradient watersheds which preserves the topology of the initial watershed lines and extracts homogeneous objects of a larger scale. The *waterfall* algorithm [15,16] is used here for producing a nested hierarchy of partitions, $P^h = \{r_1^h, r_2^h, ..., r_{m_h}^h\}; h = 1, \cdots n$, which preserves the inclusion relationship $P^h \supseteq P^{h-1}$, implying that each atom of the set P^h is a disjoint union of atoms from the set P^{h-1}. For successively creating hierarchical partitions, the waterfall algorithm removes from the current partition (hierarchical level) all the boundaries completely surrounded by higher boundaries (see Figure 4). Thus, the saliency of a boundary is measured with respect to its neighborhood. The iteration of the waterfall algorithm ends with a partition of only one region.

In our implementation of the waterfall, the saliency measure of a boundary is based on a collection of energy functions used to characterize desired single-

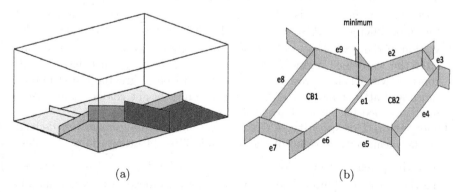

(a)   (b)

Fig. 4. (a) Partition with valuated frontiers and (b) example of region boundary: as the value of edge e1 is smaller than the values of its neighboring edges (e2 to e9), it will be removed by the waterfall algorithm

region properties and pair-wise region properties. The single region properties include region area, region convexity, region compactness and color variances within the region. The pair-wise properties include color mean differences between two regions and edge strength.

Using these energy functions as region merging criteria, the saliency driven perceptual grouping process (section 4) results in the formation of Object Of Interest (OOI). The proposed method allows multiple OOIs to be segmented according to the saliency map, by incorporating an inhibition of return (IOR) mechanism (section 5), which resets the selected OOI.

Conform the previous discussion about object-based attention, the AR corresponds to an early formed proto-object, while the OOI corresponds to a reshaped hierarchical proto-object, comprised of several lower-level hierarchically organized proto-objects.

3   Saliency Map

In the second stage of the attentional selection model, the segmented regions are integrated together, in a competitive manner, into a saliency map $\mathcal{S}$ in accordance with equation (1) representing the saliency of a region R_i at the current hierarchical level h. A region's saliency is determined by its position factor, the sum of its contrast compared with the neighboring regions, its shape and orientation.

$$\mathcal{S}(R_i) = \frac{CSR(R_i)OC(R_i)SI(R_i)}{CBI(R_i)} \tag{1}$$

where,

- **Contrast:** $CSR(R_i)$ is the *normalized mean color contrast* of a region with the surrounding regions, defined as $CSR(R_i) = \sum_{j=0}^{N-1} \alpha_{i,j}$ $(\sqrt{(\Delta\mu_L(R_i,R_j))^2+(\Delta\mu_a(R_i,R_j))^2+(\Delta\mu_b(R_i,R_j))^2}-T_d)$, N the number of adjacent regions of R_i. $\alpha_{i,j}$ is the ratio of the length of the common boundary of R_i and R_j, over the perimeter of R_i ($\alpha_{i,j} = \frac{Length(\delta R_i \cap \delta R_j)}{Perimeter(R_i)}$). The normalization factor T_d is estimated as $T_d = \mu_d - \sigma_v$, with μ_d the mean of the color differences D_i's, and $\sigma_v = \sqrt{1/n \sum_{i=1}^{n}(D_i - \mu_d)^2}$ the standard deviation of the $n = \frac{k(k-1)}{2}$ color differences between the k generated regions after the watershed segmentation [17]. Indeed, regions, which have a high contrast with their surroundings, are likely to be of greater visual importance and attract more attention. For instance, bright colors set against a more subdued background are likely to draw the eye.
- **Orientation Conspicuity:** $OC(R_i)$ is the *orientation conspicuity* defined as the mean output value of the steerable filter (4 orientations, 3 scales) over the pixels in the region R_i, $Area(R_i)$; $OC(R_i) = \frac{\sum_{p \in R_i} \hat{O}_p}{Area(R_i)}$; $\hat{O}_p$ being the normalized orientation map (at pixel p). Indeed, orientation map is an important recognition cue, here, it is also employed to describe region orientation information importance, and calculated as defined in [10] (see section 2).

- **Shape Indicator:** $SI(R_i)$ is a *shape indicator* expressing the compactness of the region, defined as $SI(R_i) = \frac{perimeter(R_i)}{Area(R_i)}$. With this parameter, we try to find a trade-off between articulated regions and more compact regions of different sizes.

- **Compositional Balance Indicator:** $CBI(R_i)$ is the *compositional balance indicator* [14]. Let $gc(R_i)$ be the center of gravity of region R_i; $gc(R)$ the gravitational center of all regions in the image with respect to their saliency value and size, defined as $gc(R) = \frac{\sum^{regions} S(R_i) Area(R_i) gc(R_i)}{\sum^{regions} S(R_i) Area(R_i)}$; R' the region whose gravitational center is the nearest neighbor of the symmetrical point of $gc(R_i)$ with respect to the midline of the image, this as a measure of overall content of the image. Then,

$$CBI(R_i) = \begin{cases} ||gc(R_i) - gc(R)||; gc(R) \in R_i \\ ||CSR(R_i)|| + ||CSR(R')|| + ||gc(R_i) - gc(R)||; otherwise \end{cases}$$

If the salient region is located near $gc(R)$, we know that the larger CSR and the nearer distance between its gravitational center and the $gc(R)$ region in the image is, the smaller CBI of the region is, meaning the higher the possibility that it will be a salient portion of the image frame. For the second case, the higher CBI (high $CSR(R_i)$ and high $CSR(R')$) shows that the image frame may balance two or more elements encouraging our eye moving between these regions. If $CSR(R_i)$ is high and $CSR(R')$ is low, than CBI will be lower, resulting in a higher saliency compared to the previous described situation, where both $CSR(R_i)$ and $CSR(R')$ are high.

As such, saliency is guided by the overall content of the image, represented inherently by the CBI factor. Note that CBI depends on the saliency of the regions ($gc(R)$). Therefore the initialization of CBI uses the saliency value of all regions, calculated as in Equation 1, omitting the CBI factor.

The region with the highest saliency value $\mathcal{S}(R_i)$ is selected as attention region (AR).

4   Perceptual Grouping

The pivotal idea in the proposed model of object-based visual attention is, in conformity with psychophysical findings, grouping based salience computation, salience driven perceptual organization and integrated competition for focus of attention. Perceptual grouping of segmented regions is envisaged, starting from the salient region, to obtain perceptually meaningful regions. The salience of a grouping measures how different this grouping contrasts with its surroundings and depends on various factors, such as feature properties, perceptual grouping, dissimilarity between the attention region (AR) and its neighborhood, etc.

After the selection of the AR, a perceptual grouping process is applied using the waterfall algorithm [15], [16] based on a saliency attached to each edge of the RAG. As such, preserving the topology of the initial watershed transformation

on the combined gradient image, a nested hierachy of partitions is obtained. A
grouping is considered a hierarchical structure of objects and space [2].

The region saliency mappings $\mathcal{S}(R_i)$ are dynamically varied according to competition conditions among the groupings at different hierarchical levels of the waterfall. In our implementation of the waterfall, the saliency measure of a boundary is based on a collection of energy functions used to characterize desired
single-region properties and pair-wise region properties, following the formulation of [18]. The single region properties include region area, region convexity,
region compactness and color variances within the region. The pair-wise properties include color mean differences between two regions and edge strength along
the shared boundary. The saliency of the boundary between two neighboring
regions R_i and R_j:

$$E(\tilde{R} = R_i \bigcup R_j | R_i, R_j) = E(\tilde{R}) + E(R_i, R_j) \tag{2}$$

Where $E(\tilde{R} = R_i \bigcup R_j | R_i, R_j)$ is the cost of merging the regions R_i and R_j, $E(\tilde{R})$
is the merged region property (saliency) and $E(R_i, R_j)$ the pair-wise property,
respectively defined as follows.

$$E(\tilde{R}) = E_{area}(\tilde{R}) \frac{1}{E_{hom}(\tilde{R})} \sum_c E_{var_c}(\tilde{R})$$
$$(1 + |E_{conv}(\tilde{R})|)^{sign(E_{conv}(\tilde{R}))} (1 + |E_{comp}(\tilde{R})|)^{sign(E_{comp}(\tilde{R}))} \tag{3}$$

$$E(R_i, R_j) = E_{edge}(R_i, R_j) E_{CMDif}(R_i, R_j) \tag{4}$$

The factors in equations 3 and 4 need to be normalized and are defined as:

- **Area**

$$E_{area}(R_i) = 0.002 \frac{NM}{Area(R_i)}$$

NM being the image size. Larger area is always preferred. The normalization
factor 0.002 means a region that has a 0.2% area of the whole image will have
an energy of 1.0. A zero-area region has an infinite area energy value and a
whole image region has an area energy value of 0.002. A penalty function is
used to prevent to large regions. Typically regions are penalized when there
area is greater than ±25% of the whole image area.

- **Convexity**

$$E_{conv}(R_i) = \frac{Area(ConvexHull(R_i))}{Area(R_i)} - 1.25$$

represents the region convexity energy. We assume that regions with convexity larger than 1.25 are not preferred, and those with convexity energy
smaller than 1.25 are desired. Therefore, the offset for the convexity energy
is set to -1.25.

- **Compactness**

$$E_{comp}(R_i) = \frac{Perimeter(R_i)^2}{4\pi Area(R_i)} - 1.25$$

represents the region compactness energy.

The compactness energy $\frac{Perimeter(R_i)^2}{4\pi Area(R_i)}$ is always greater than or equal to 1 (1 for a circle, $4/\pi$ for a square). We assume that regions with compactness larger than 1.25 are not preferred. Again the offset for compactness energy is set to -1.25.

- **Homogeneity**

$$E_{hom}(R_i) = 1 - \sigma(R_i)V(R_i)$$

represents the region's intensity (I) homogeneity. Homogeneity is largely related to the local information extracted from an image and reflects how uniform a region is. In [17], an interesting pixel-based homogeneity definition is presented, as a composition of two components: standard deviation and discontinuity of intensity I. Standard deviation describes the contrast within a local region. Discontinuity is a measure of abrupt changes in gray levels and could be obtained by applying edge detectors to the corresponding region.

− **Color Variance**

$$E_{var_c}(R_i) = \frac{1}{15}\sigma_c(R_i)$$

represents the color homogeneity of a region, with $\sigma_c(R_i)$ the standard deviation of the color $c \in \{L, a, b\}$ within region R_i. The normalization factor for color variances is derived from statistical analysis of the color variance results on image data base [18].

− **Color Contrast**

$$E_{CMDif}(R_i, R_j) = \sqrt{(\Delta\mu_L(R_i, R_j))^2 + (\Delta\mu_a(R_i, R_j))^2 + (\Delta\mu_b(R_i, R_j))^2} - T_d$$

represents the normalized mean color difference. The normalization factor T_d defined previously in section 3. Following the proposal in [17], The normalization factor T_d (for color merging) is estimated as $T_d = \mu_d - \sigma_v$, with μ_d the mean of the color differences D_i's, and $sigma_v = \sqrt{1/n \sum_{i=1}^{n}(D_i - \mu_d)^2}$ the standard deviation of the $n = \frac{k(k-1)}{2}$ color differences between the k regions [17].

Using these energy functions as region merging criteria, the saliency driven perceptual grouping process results in the formation of Object Of Interest (OOI).

5   Iterative Object Popping out Process − Inhibition of Return

A region is selected as Object Of Interest (OOI) when the local region merging process is completed. The attempts to merge are performed in the neighborhood of the AR. When no more merges are allowed in this neighborhood, following the energy criteria, the local merging is completed.

The proposed method allows multiple OOIs to be segmented according to the saliency map, by incorporating an inhibition of return (IOR) mechanism, which resets the selected OOI and restarts the entire procedure by selecting the new most salient AR.

The selection procedure of the AR consists of ordering the segments in terms of highest saliency value $\mathcal{S}(R_i)$, instead of using a *Winner-Take-All* (WTA) network as proposed by [19] and the NVT [20]. Although biologically less plausible, equivalent results are achieved with less computational resources. The segment with highest saliency value is selected as attention region (AR). In contrast with existing state of the art approaches ([10], [8]), we obtain directly the AR based on our region saliency computation.

Once an OOI is focused on, another computational problem is posed: how can we prevent attention from permanently focusing onto the same OOI? In [21], one efficient computational strategy is proposed, consisting of suppressing the currently attended location in the saliency map. Hence, the winner-take-all network naturally converges towards the next most salient location. Repeating this process generates attentional scanpaths. Such inhibitory tagging of recently attended locations has been widely observed in human psychophysics as a phenomenon called 'inhibition of return' (IOR) [22].

Computationally, IOR implements a short-term memory of the previously visited locations and allows the attentional selection mechanism to focus instead on new locations. The simplest implementation of IOR, suppressing the attended location, only represents a coarse approximation of biological IOR, which has been shown to be object bound.

In the proposed model of object-based visual attention, we are only concerned with covert attention, that is, shifts of the focus of attention in the absence of eye movements. Although simple in principle, IOR is computationally a very important component of attention, in that it allows us, or a model, to rapidly shift the attentional focus over different items (OOIs) with decreasing saliency, rather than being bound to attend only to the OOI of maximal saliency at any given time.

Although pop-out detection and IOR are named as two different processes, they are very much interdependent on each other. The IOR greatly influences the process of pop-out by dictating what not to attend in the consequent attention cycle. In general, two types of inhibitions are considered, top-down and bottom-up. The top-down influence is regarded as an external stimulus coming from long term knowledge, recent experiences, and current needs. The other type of inhibition occurs within the attention mechanism to avoid repeatedly focusing on the same object. The top-down influence is not included in our implementation.

It has been established by experiments in psychophysics that inhibition takes place in terms of both location and object features [23]. Evidence is provided for inhibition in the immediate vicinity of the attended location and a U-shaped function has been reported which strongly suppresses the immediate surroundings of the attended location and gradually fades to no suppression after a limited diameter.

In general, IOR has single influence directly on the resulting saliency map, meaning the saliency map will not change apart from the inhibited focused region

([20], [10]). In contrast, we consider two types of inhibition mechanisms, namely spatial based and feature based. Introducing IOR at lower levels of the computation (as suggested in [24]), implies the evolution of the region saliency value of all segments. The spatial inhibition inhibits the region that is selected as an attention region (AR), and excludes this region to be refocussed on. The feature based inhibition inhibits the influence of the focussed region on its neighborhood regions, in terms of similarity measure calculation between two regions to obtain the cost-value of the arcs, as well as saliency calculation of the remaining, unfocussed regions. From an implementation point of view, this is achieved by simply excluding the focussed region from the RAG. We formulate these factors as:

$$S'(R_i) = \xi_s \frac{CSR(R_i, \xi_f)OC(R_i)SI(R_i)}{CBI(R_i)} \tag{5}$$

$$\xi_s = \begin{cases} 0; R_i == OOI \\ 1; otherwise \end{cases} \tag{6}$$

where ξ_f excludes the OOI from CSR calculation.

The human browsing behavior can be approximately modeled by two mutually exclusive statuses: the fixation status (e.g., exploiting an interesting region) and the shifting status (e.g., covertly scrolling to the next region). The fixation status corresponds to the static viewing of an attention objects, and the shifting status can be simulated by covertly traveling between different attention objects. The shifting path is the shortest path between centers of the two fixation areas (i.e. objects of interest). The iterations of these two states compose the whole simulation of the shifting process, forming a scan path. The saccade status can be described as a shifting process from the most informative region to the second one, then the third and so on.

The scan path starts with the pop-out of the most informative object of interest. is formed by subsequently focussing on OOIs in the image. In [25], *minimal perceptible time (MPT)* is introduced as a threshold for the fixation duration when focusing on an OOI. If an attention object does not stay on the screen longer than MPT, it may not be perceptible enough to let users catch the information. Fixation durations are variable, typically ranging from 100 ms to 500 ms. The MPT of an OOI is proportional to its region saliency value.

Subsequently, the attended OOI is inhibited. The inhibition of return mechanism works as a short term memory, and stores attended OOIs in the inhibition map. Taking into account information from the inhibition map, the region features are updated. Hence, the region saliency map is updated as well as the edge energy values.

Since humans are known to make fixations on nearby objects, the saliency map (saliency value of each region) is weighted by the proximity to the current fixation, defined as

$$w_{saccade,i} = \frac{1}{\sqrt{(gc(OOI) - gc(R_i))^2}} \tag{7}$$

with $gc(OOI)$ the gravitational center of the selected Object of Interest region; $gc(R_i)$ the gravitational center of any remaining region. The weighted saliency map is then calculated as

$$S^w(R_i) = w_{saccade,i}S(R_i) \tag{8}$$

with $S(R_i)$ as defined in Equation 1. The scan path determination process continues with a new AR: the newly highest salient region excluding the already detected OOI's. The scan path is complete when (a part of) the background is selected, or when there are no more meaningful regions left.

6   Experimental Results

We have tested our computational model for object-based attention on natural scene images, where a set of objects distinguishes from the background. We attempt to determine the scan path, by subsequently focussing our attention on Objects of Interest. As discussed in the previous sections, in a first phase, features are calculated with respect to small regions, rather than pixel based. This approach is illustrated in Figure 5. From these feature maps, a combined gradient is estimated using the method described in [13], with the purpose of eliminating misleading high-contrast edges.

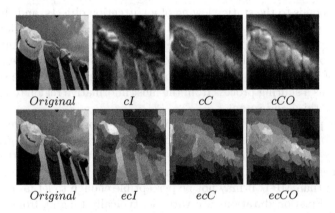

| Original | cI | cC | cCO |

| Original | ecI | ecC | ecCO |

Fig. 5. Comparison of conspicuity maps (c) and the proposed enhanced conspicuity maps (ec), incorporating region information. The presented maps represent intensity (I), orientation (O), color (C), and color opponency (CO)

The results of the scan path determination are presented in Figure 6 and in Figure 7. The top left images depict the original image together with the starting mosaic image. The rest of the figures illustrate the scan path. For each Object of Interest that has gained the focus of attention, the following resulting maps are presented:

– Mosaic Image, corresponding to the current merge results in the image (hierarchical level).

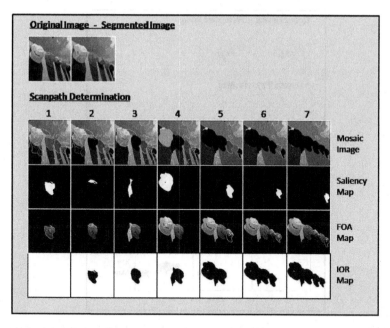

Fig. 6. Hats-image scan path determination. For each Object of Interest that has gained the focus of attention, the corresponding Mosaic Image, Saliency Map, Focus of Attention (FOA) Map, and Inhibition of Return (IOR) Map are presented.

- Saliency Map, displaying the saliency value (normalized between [0-255]) of each region at the moment when the scan path is updated.
- Focus of Attention (FOA) Map, representing the scan path. The current Object of Interest is added to the focus of attention map, and highlighted by means of white edges and the index in the scan sequence of Objects of Interest.
- Inhibition of Return (IOR) Map. Whenever an Object of Interest is added to the scan path, the IOR Map is updated after extraction of the OOI. More concretely, the first OOI extracted in step 1 of the scan path development, is inhibited during the subsequent scan path augmentation, and stays inhibited until the scan path is complete.

In Figure 8, a comparison in terms of scan path determination is performed between the presented approach of Object of Interest (OOI) scan path determination and the method for salient region extraction using the publicly available Saliency Toolbox ([8]). Parameters for the experiment using the Saliency Toolbox are: equal weights for all features (color, intensities, 4 orientations), lowest surround level 3, highest surround level 5, smallest c-s delta 3, largest cs-delta 4, saliency map level 3, iterative normalization with 3 iterations. The scan path determined with the Saliency Toolbox selects the most salient regions iteratively, however the method does not consider the real object borders. In the presented

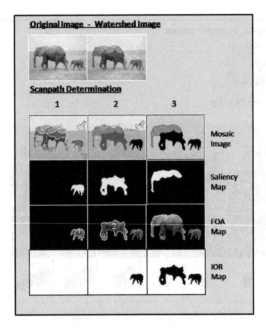

Fig. 7. Elephants-image scan path determination. For each Object of Interest that has gained the focus of attention, the corresponding Mosaic Image, Saliency Map, Focus of Attention (FOA) Map, and Inhibition of Return (IOR) Map are presented.

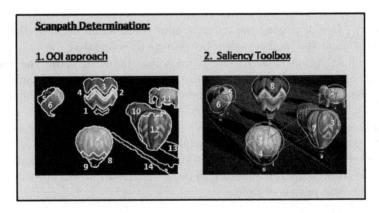

Fig. 8. Comparison of the presented approach of Object of Interest (OOI) scan path determination to the existing Saliency Toolbox ([8]). The subsequent focussed regions in the scan path are numbered.

approach, we try to extract objects of interest with exact localization of object borders.

We can conclude that our approach gives excellent results in extracting meaningful, perceptual objects when objects have limited intern color differences.

However, when large color differences occur between object parts, objects are selected in several steps.

7 Conclusions

This work reports a novel region-based attention model. Unlike existing models, the proposed approach performs segmentation in meaningful regions as an attentive process, during which only visually salient image regions are merged using perceptual organization criteria. In addition, the proposed approach involves a multi-scale concept allowing the segmentation of meaningful regions at the corresponding scale, using the waterfall algorithm. Experimental results demonstrated the usefulness of the approach.

Three extensions of the method will be investigated in future work. First, other segmentation techniques, such as mean shift segmentation, will be used to reduce the original number of regions in the image. Second, extending the idea of Attention Region (AR) to Attention Window (AW) which could be created based on the region saliency map and saliency points from an image. Using such AW will allow a better extraction of OOIs using the proposed region merging criteria. Finally, the current bottom-up region-based model of visual attention needs extension towards top-down (endogenous) attentional processes. Following the suggested clarifications on the inconsistencies in existing models for visual attention, particularly concerning the interaction between bottom-up and top-down processes [26], [27], the discussion has been started in cooperation with the department of Psychology and Educational Sciences.

References

1. Itti, L.: Models of Bottom-Up Attention and Saliency. In: Itti, L., Rees, G., Tsotsos, J.K. (eds.) Neurobiology of Attention, pp. 576–582. Elsevier, Amsterdam (2005)
2. Sun, Y.: Hierarchical Object-Based Visual Attention for Machine Vision. PhD. Dissertation, University of Edinburgh (2003)
3. Jarmasz, J.: Towards the Integration of Perceptual Organization and Visual Attention: The Inferential Attentional Allocation Model. PhD. Dissertation, Carleton University (2001)
4. Treisman, A., Gelade, G.: A feature Integration Theory of Attention. Cognitive Psychology 12, 97–136 (1982)
5. Sun, Y., Fisher, R.: Object-based Visual Attention for Computer Vision. Artificial Intelligence, 77–123 (2003)
6. Scholl, B.J.: Objects and Attention: the state of the art. Cognition 80(1-2), 1–46 (2001)
7. Walther, D., Rutishauser, U., Koch, C., Perona, P.: Selective visual attention enables learning and recognition of multiple objects in cluttered scenes. Computer Vision and Image Understanding 100, 41–63 (2005)
8. Walther, D., Koch, C.: Modeling attention to salient proto-objects. Neural Networks 19, 1395–1407 (2006)

9. Fritz, G., Seifert, C., Paletta, L., Bischof, H.: Attentive object detection using an information theoretic saliency measure. In: Paletta, L., Tsotsos, J.K., Rome, E., Humphreys, G.W. (eds.) WAPCV 2004. LNCS, vol. 3368, pp. 29–41. Springer, Heidelberg (2005)

10. Fintrop, S.: VOCUS: A Visual Attention System for Object Detection and Goal Direction. LNCS (LNAI), vol. 3899. Springer, Heidelberg (2006)

11. Liu, H., Jiang, S., Huang, Q., Xu, C., Gao, W.: Region-Based Visual Attention Analysis with Its Application in Image Browsing on Small Displays. In: MM 2007, pp. 305–308 (2007)

12. Meyer, F.: An overview of Morphological Segmentation. IJPRAI 15(7), 1089–1118 (2001)

13. O'Callaghan, R.J., Bull, D.R.: Combined Morphological-Spectral Unsupervised Image Segmentation. Image Processing 14(1), 49–62 (2005)

14. Congyan, L., De, X., Xu, Y.: Perception-Oriented Prominent Region Detection in Video Sequences. Informatica 29, 253–260 (2005)

15. Beucher, S.: Watershed, hierarchical segmentation and waterfall algorithm. In: Beucher, S. (ed.) Mathematical Morphology and its Applications to Image Processing. In: Proc. ISMM 1994, pp. 69–76 (1994)

16. Marcotegui, B., Beucher, S.: Fast implementation of waterfall based on graphs. In: Proc. 7th international symposium on mathematical morphology, pp. 177–186 (2005)

17. Cheng, H.-D., Sun, Y.: A hierarchical approach to color image segmentation using homogeneity. IEEE Trans. on Image Processing 9(12), 2071–2082 (2000)

18. Luo, J., Guo, C.: Perceptual grouping of segmented regions in color images. Pattern Recognition 36, 2781–2792 (2003)

19. Koch, C., Ullman, S.: Shifts in selective visual attention: towards the underlying neural circuitry. Human Neurobiology 4, 219–227 (1985)

20. Itti, L., Koch, C., Niebur, E.: A model of saliency-based visual attention for rapid scene analysis. IEEE Trans. on Pattern Analysis and Machine Intelligence 20, 1254–1259 (1998)

21. Itti, L., Koch, C.: Computational modelling of visual attention. Nature Reviews Neuroscience 2(3), 194–203 (2001)

22. Klein, R.: Inhibition of return. Trends Cogn. Sci. 4, 138–147 (2000)

23. Aziz, M.Z., Mertsching, B.: Pop-out and IOR in Static Scenes with Region Based Visual Attention. In: The 5th International Conference on Computer Vision Systems (ICVS) (2007)

24. Henderickx, D., Maetens, K., Soetens, E.: Inhibition of return: A bottom-up routed attentional process (submitted, 2008)

25. Xie, X., Liu, H., Ma, W.-Y., Zhang, H.-J.: Browsing large pictures under limited display sizes. IEEE Trans. on Multimedia 8(4), 707–715 (2006)

26. Henderickx, D., Maetens, K., Soetens, E.: Understanding the Interactions of Bottom-up and Top-down Attention for the Development of a Humanoid Robot System. In: Paletta, L., Tsotsos, J.K. (eds.) Proceedings of the Fifth International Workshop on Attention and Performance in Computational Vision, Santorini, Greece, pp. 124–137 (May 2008)

27. Henderickx, D., Maetens, K., Geerinck, T., Soetens, E.: Modeling the interactions of bottom-up and top-down guidance in visual attention. In: Paletta, L., Tsotsos, J.K. (eds.) WAPCV 2008. LNCS (LNAI), vol. 5395. Springer, Heidelberg (2008)

Attention Mechanisms in the CHREST Cognitive Architecture

Peter C.R. Lane[1], Fernand Gobet[2], and Richard Ll. Smith[3]

[1] School of Computer Science, University of Hertfordshire
`peter.lane@bcs.org.uk`
[2] School of Social Sciences, Brunel University
`fernand.gobet@brunel.ac.uk`
[3] School of Social Sciences, Brunel University
`richard.smith@brunel.ac.uk`

Abstract. In this paper, we describe the attention mechanisms in CHREST, a computational architecture of human visual expertise. CHREST organises information acquired by direct experience from the world in the form of *chunks*. These chunks are searched for, and verified, by a unique set of heuristics, comprising the attention mechanism. We explain how the attention mechanism combines bottom-up and top-down heuristics from internal and external sources of information. We describe some experimental evidence demonstrating the correspondence of CHREST's perceptual mechanisms with those of human subjects. Finally, we discuss how visual attention can play an important role in actions carried out by human experts in domains such as chess.

1 Introduction

Cognitive science studies the processes by which humans develop and manifest intelligent behaviour. The study of visual perception has been widely recognised as an important component of many areas of expertise. A seminal experiment by de Groot [1] uncovered a central component of human expertise: the ability to identify the important features of a stimulus in the domain of expertise. De Groot's experiments were performed on chess players, and involved a test of *recall ability*. Each participant was shown a position on a chess board containing approximately 23 pieces for a few seconds, and was then asked to reconstruct the position from memory. Candidates for the world title managed to reconstruct the position with few, if any, errors; average players managed much worse. As there was no difference in the participants' intelligence level, other visual skills or general memory, clearly the difference was related to their level of expertise.

An explanation for the difference had to await the development of the chunking theory [2,3] and cognitive models of human learning such as EPAM (Elementary Perceiver And Memoriser) [4]; the most detailed model of chess expertise is now CHREST (Chunk Hierarchy REtrieval STructures) [5,6,7,8], which is a version of EPAM with the addition of templates and more sophisticated perceptual mechanisms. The computational modelling and experiments have shown that,

L. Paletta and J.K. Tsotsos (Eds.): WAPCV 2008, LNAI 5395, pp. 183–196, 2009.

essentially, the difference in perceptual skills can be explained by two factors: the development of a large (approximately 300,000 chunks) set of knowledge about the domain, and the use of templates by the visual system to actively seek out higher-order clusters of information in a stimulus.

We can summarise the role of attention within CHREST as follows. First, an image is perceived, and, as the eye has a limited field of view, a portion of the image has its features extracted. These features are then used to sort through long-term memory, seeking a familiar pattern. Any retrieved pattern is placed into short-term memory. The contents of short-term memory, some high-level domain-specific knowledge, and any items on the periphery of the field-of-view will all combine to guide the model's eye to locate a new point of the image to focus on. This process continues, and the model will attempt to build up, in short-term memory, a set of pointers to familiar patterns in long-term memory which 'cover' the image. Here we see the importance of prior experience, as the capacity of short-term memory is limited, but experts will have larger familiar patterns (chunks), and so can store much more relevant information in their short-term memory. However, attention is also important, as the larger patterns must be confirmed to be present in the image if they are to be maintained in short-term memory. Later, we explain in detail how a set of heuristics manages this process in a dynamic and flexible manner.

This paper continues with an overview of the CHREST architecture, details of the attention mechanism, and summaries of some results from experiments using CHREST to explore visual abilities. Finally, we discuss how the findings with CHREST relate to current issues in the study of attention.

2   Overview of CHREST

CHREST is an example of a *cognitive architecture*, that is, an implementation of a theory of human cognition for developing detailed models of human behaviour in a range of domains. The strength of cognitive architectures as scientific theories is that their implementation ensures a high degree of precision in the theory's formulation, providing a sufficiency proof that the proposed mechanisms can carry out the tasks in the domain of interest. Analysis of how the theory's predictions match actual behaviour, using measures such as eye movements, reaction times, and error patterns, also establish the quality of the cognitive architecture, by confirming its behaviour against actual human data.

Together with EPAM, from which it is derived, the CHREST architecture has been a source of successful models of human learning and perception over a 50-year period [3]. Beyond the results discussed in this paper, these successes have included results in: verbal learning, language acquisition [9], categorisation and problem solving. In this section, we provide an overview of CHREST, explain its key processes, and highlight some important empirical results. The next section will focus in more detail on the important attention mechanisms.

Fig. 1 shows the principal components of CHREST. The architecture follows a classic subdivision of the cognitive processes into three components: one of

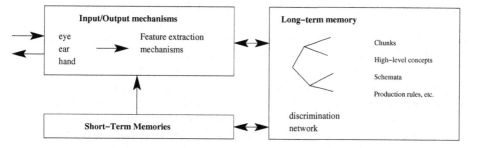

Fig. 1. The CHREST cognitive architecture

input/output and two of memory. The input/output mechanisms are used to interact with the environment. Input stimuli are separated into features, and output actions are converted into motor controls. The Long-Term Memory (LTM) is used to store information in a more-or-less permanent state; all information learnt from the environment is held in LTM. The LTM uses a discrimination network, which is constructed incrementally during the system's lifetime, to index a pool of familiar patterns (chunks), higher-level concepts, schemata and productions. The final component of the architecture is the Short-Term Memory (STM), which provides a temporary store for information retrieved from LTM whilst considering a particular stimulus or problem. Different parts of STM relate to different modalities of information, such as visual or verbal. The STM maintains an *hypothesis node*, which is the most informative node retrieved so far, and is used to focus CHREST on retrieving or constructing large chunks.

CHREST is an example of a symbolic cognitive architecture, which means that all information retrieved by and stored within a CHREST model is at a high level. We shall use the term 'patterns' and 'chunks' throughout this paper. A pattern is what CHREST perceives in the external world, and a chunk is a familiar pattern, one retrieved from its internal memory. Both patterns and chunks are represented as lists of items with their positions. For example, in chess, the items are actual pieces, and the positions are squares on the chess board, so (Kg1 Rf1 Ph2 Pg2 Pf2) would describe a typical castled position for the White player.

3   CHREST's LTM: Chunking Networks

CHREST's long-term memory is a *chunking network*: a discrimination-network representation with cognitively plausible learning mechanisms, based on the chunking theory. The chunking network includes mechanisms for forming and using *lateral links*, as well as clustering techniques, to form *templates*.

3.1   Learning and Retrieving Perceptual Chunks

A *chunk* is a familiar pattern: CHREST acquires and retrieves chunks from its long-term memory through a set of learning and storing operations acting on a discrimination network. The network is formed from *nodes* holding the chunks; the chunk stored at a node is known as its *image*. There is a *root node* from which all sorting operations begin, and the discrimination network is built up from the *test links* between nodes. Patterns are sorted through the network by following those test links whose tests match the given pattern.

The network is constructed using two learning mechanisms. *Discrimination* is used when a pattern is sorted to a node which has no succeeding test links matching the pattern, and the pattern mismatches the chunk stored at the node. The part of the pattern which mismatches the chunk is used as the test for a new test link. *Familiarisation* occurs when a pattern is sorted to a node which has no succeeding test links matching the pattern, and the pattern matches the chunk stored at the node. If the pattern contains more information than is held at the node, then extra information is added to the stored chunk. Thus, node images are specialised as patterns are re-encountered.

In describing the discrimination and familiarisation operations, part of the sorted pattern is used, either as a new test, or to augment an existing chunk. The amount of the sorted pattern used is a function of the rest of the network in the following sense. The part of the pattern to be used is *re-sorted* in the network, and the chunk that is retrieved is used in the above processes. This chunking process ensures that learning is slow in the initial stages of getting to know a domain, but as the network grows, and some of the patterns become known, the rate of learning will speed up.

Efficiency of accessing a large store of data is often a concern, as in the utility problem [10]. However, the usable capacity of a data structure need not be problematic if it is sufficient in the given domain [11]. Human experts typically pick out familiar patterns in a time of around 250ms, which is easily matched by CHREST, using a network of around 300,000 chunks [8].

3.2   Constructing Templates

The *template* is a critical element of CHREST's explanation of the recall abilities of human experts [8]. A template collects together the information stored in a number of separate nodes so as to highlight the constant *core* information and the variable information, held in *slots*.

Template creation is assumed to occur whenever a critical condition is met by a specific node within the network. Similarity between nodes is made explicit by providing *similarity links* between them (such links are discussed in the next subsection). When a node has a sufficient number of such links, and the contents of these nodes satisfy an overlap criterion, the information in these nodes is aggregated to form a template. Work in chess [8] has assumed a threshold for the number of similarity links of 4, with at least 5 features in common.

3.3   Lateral Links

Lateral links [12] are created when the model has retrieved two chunks (retrieved chunks are stored in STM); two classes of lateral link may be distinguished.

1. Clears the record of which heuristic was last used.
2. Selects a point in the centre of the field of view to start from.
3. Performs following cycle:
   (a) Store the currently fixated item in the list 'fixated-items'
   (b) Gets the next fixation point using `get-next-fixation` (see text).
4. If timing is used and model has used its presentation time, or if the maximum number of allowed fixations is reached, then finish.
5. Otherwise, repeat from step 3.

Fig. 2. CHREST's perception cycle; learning occurs during `get-next-fixation`

The first is where the two nodes match a similarity function of some kind. The simplest of these is the direct *similarity link*, as used above in the creation of templates. More complex *generative links* [9] connect those nodes with similar descendant test links. The second class of links are those where the chunks are of different type. For instance, *production links* are used to associate a perceptual pattern with its corresponding action or conclusion, and so the production link forms a basis for the model's problem-solving behaviour. Gobet et al. [3] provides a summary of these links and their use in different domains; Section 5.2 provides more details on how such links are formed.

4   CHREST's Attention Mechanism

CHREST uses a simulated eye to retrieve information from its target stimulus. The eye is directed to a focus of attention, the *fixation point*, and has a limited field of view. Its movement is governed by a set of heuristics, which combine low-level and high-level information. A *perception-learning-perception* cycle guides CHREST's eye movements around the current stimulus for the presentation time; see Fig. 2. In the following, we describe the heuristics and perception processes to be introduced in the latest release of CHREST, version 3.0.

The list of 'fixated-items' is part of CHREST's short-term memory. The fixated-items record the items and their positions fixated upon during the current presentation cycle. Each item observed will be added to this list until a termination condition is met. This termination condition is when an empty location is fixated, or CHREST fixates something already in the list, or the heuristic used is a random or weak heuristic. Marking the list as fixated will then lead to the fixated-items list being learnt as a pattern. After learning, the fixated-items list is cleared, and CHREST will begin building a new list, as it continues its perception cycle. Worth noting is the implicit connection between perception and learning; the list of fixated items is constructed by sequences of eye fixations, which means that information learnt by CHREST has a locality bias.

Heuristics for selecting eye fixations can come from two sources. First, there are the generic, domain-neutral heuristics which are part of the general architecture. These heuristics include: using LTM, fixating a part of the scene not observed yet, locating a random object on the periphery, and locating a random position on the periphery. Second, there are domain-specific heuristics which are part of the model for that domain alone. For example, in the chess models there is an heuristic to guide the eye to typical positions of the kings; attacking or defending the king is the ultimate key to winning at chess, and so the king position is an important initial factor in assessing a position. Again in chess, there is an heuristic to guide the eye to a square attacked by the currently fixated piece.

The process `get-next-fixation` is responsible for applying the various heuristics to retrieve a new fixation point. The process uses a hierarchy of heuristics: first it tries to use LTM, second it tries fixating a new object or a position suggested by the domain-specific heuristics, finally it defaults to choosing a new random position. There is a stochastic element, which means CHREST will fail to use an applicable heuristic in a percentage of cases (currently set at 10%).

As an example we describe in more detail the most interesting heuristic, which selects a new square based on information in LTM. This heuristic uses pointers into long-term memory as a basis for selecting the next fixation point. There are two main sources. The first is the *hypothesis node*, held in short-term memory, which acts as an anchor, guiding CHREST to retrieve or construct the largest chunk possible for the given stimulus. The test links of the hypothesis node (its descendants) are considered in turn, as described below. The second is the node retrieved by learning something from the currently perceived scene. The process is as follows: (1) `current-node` is the result of learning from the current scene; (2) if there are any remaining descendants of the hypothesis to consider, let `current-child` be the first descendant, and remove it from the remaining list; (this step attempts to retrieve the largest chunk) (3) otherwise, let `current-child` be the first child-link of the `current-node`; (4) return the first potential square of `current-child` as the next fixation point. The learning that can occur in step (1) attempts to extend the hypothesis with information currently perceived, or else using the list of fixated-items, if that list has been marked as complete; hence, CHREST is biased towards learning the largest chunk possible.

One consequence of linking low-level perceptual processes directly with the higher-level information stored within the long-term memory is a blurring of the boundary between primitive and complex visual objects. We assume that CHREST initially contains single-element chunks for each of the primitive elements which may occur in the given domain, e.g. if CHREST is scanning text documents, all the single letters will be provided as primitive features. As CHREST learns about the domain, it begins to form larger chunks, consisting of groups of these primitive features. In recognition, CHREST will typically retrieve complete chunks from the domain, bypassing the more elementary features. Thus, in scanning text documents, CHREST soon begins to work with words as 'primitive' elements, instead of single letters. Such creation and use of higher-level features occurs in any domain, and is typical of the perceptual knowledge acquired and used by experts. It is also a typically hard problem in unsupervised learning tasks, especially perceptual domains [13,14].

5 Experiments with CHREST on Attention

We describe two sets of experiments from previous work to illustrate the impact of CHREST's perceptual mechanisms: data from chess, on the details of the human attention mechanism; and data from a word-recognition task, explaining how expectations assist in disambiguating input data.

5.1 Chess

In the introduction of this paper, we mentioned de Groot's [1] classic experiments on chess memory. It is of interest that these experiments were motivated by a need to understand chess masters' perception and attention: how can a world-class master understand more of a position after five seconds than a candidate master after fifteen minutes? How can search be so selective with strong players, who often consider no more than one hundred positions during their thinking, and ignore the billions of positions that could, in principle (but not in practice), be of interest? CHREST provides mechanisms explaining how strong players attend to the relevant while ignoring the irrelevant. These mechanisms have been used to explain both how players direct their attention when considering a position, and how they can limit the number of moves they anticipate during look-ahead search. The organisation of these two types of behaviour is hierarchical, in the sense that the pattern of eye movements is part of the explanation of selectivity at the level of move choice. In both cases, computational simulations have reproduced key aspects of human behaviour. Given space constraints, we limit our attention to three classes of phenomena simulated by CHREST.

Eye Movements as Indicators of Attention. In their analysis of players' eye movements during the brief presentation of a chess position, de Groot and Gobet [5] identified a few striking differences between novices and masters. Masters' eye movements were shorter on average than novices' (260 ms vs. 310 ms), and also showed less variability (sd = 100 ms vs. sd = 140 ms). Masters' eyes covered more squares on the board, and also covered more of the squares that were important in the position. CHREST does a very good job of simulating these data. For example, it replicates the skill effect with the average duration of eye movements (272 ms for the simulated masters vs. 315 ms for the simulated novices) and the difference in variability (97 ms vs. 154 ms). Fig. 3 shows an example set of eye fixations; what is important is how CHREST covers approximately the same amount of the important part of the board as the expert, whereas the novice player and model (not shown) cover a much smaller area. CHREST also captures the skill differences in the percentage of the board covered, and in the percentage of critical squares covered. The speeding-up of the eye movements and the increased number of important squares fixated are due to the fact that many more fixations are directed by the structure of the discrimination network with the master version of the model than with the novice version, as the discrimination network is larger with the former than with the latter.

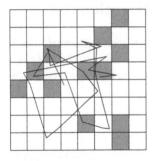   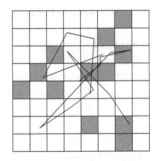

Human master   CHREST Master

Fig. 3. Example of a Master's eye movements for a specific position (left) and its simulation by CHREST (right). Grey squares indicate important squares in the position. The field of view will cover several squares around each fixation point. (After de Groot and Gobet [5].)

More qualitatively, CHREST captures masters' tendency to fixate perceptually salient pieces (e.g. a white knight on the black side of the board) early on, and their tendency not to fixate parts of the board that are "normal" in a given situation (e.g. a standard castling position).

Although we have emphasised the way perceptual knowledge helps develop more efficient attention, it should be pointed out that masters' attention is not infallible. In several of the positions used by de Groot and Gobet, immediate threats (e.g. checkmate in one move) were missed by some of the masters. The longer the distance between the two pieces, the more likely it was for the threat to be missed. Assuming that chunks play a key role in rapidly identifying threats, as we have done in this paper, this result is in line with the way CHREST learns chunks, giving precedence to relations of proximity, as described next.

The Structure of Chunks: An Archaeology of Attention. A crucial assumption in CHREST is that the information stored as chunks is a reflection of the attention mechanisms used during learning. This offers an indirect way of testing how attention is directed when playing chess. Concurrent and retrospective protocols [1,5] suggest that a fair amount of attention is directed to consideration of moves and counter moves. However, they also suggest that attention is directed to patterns of pieces. Chase and Simon [15] analysed the structure of the groups of pieces replaced together in a memory task, and found that a surprisingly small number of these groups contained relations of attack. The vast majority of these groups contained relations of proximity, same colour and defence. This result turns out to be robust, and has been replicated by [16] with a larger sample. In a simulation of the memory experiment used by Chase and Simon, Gobet [17] showed that CHREST closely simulates the detail of the pattern of relations found in chessplayers' chunks. This outcome supports the importance given by CHREST to proximity during the acquisition of chunks.

Selectivity in Choosing a Move. A characteristic of experts is the rapidity with which they can propose solutions. For example, chess grandmasters literally 'see' the good moves straight away [1,18]. With very short decision times, less than 10 seconds, their choice will not always be the best possible move, however, it will almost always be a very plausible move in the position. CHUMP (CHUnks and Moves Patterns) [19] is a variant of CHREST that implements the idea that recognising patterns of chess pieces on the board makes it possible to access information about moves in long-term memory, and thus to rapidly identify fairly good moves. CHUMP uses two different but linked discrimination networks to store two types of knowledge: first, patterns of pieces (the kind of chunks acquired by CHREST), and, second, moves and sequences of moves (see the next section for more on multi-modal learning in CHREST). During learning, the program scans positions taken from master games, and patterns of pieces are associated with moves. When selecting a move in a position in the test phase, patterns of pieces act as conditions, and moves as actions. If different patterns suggest a move, and/or if the same patterns suggest different moves, the conflict is resolved by using a function combining the number of different chunks voting for a given move and the number of times the move has been associated with a given pattern.

CHUMP provides a demonstration that the idea of selectivity through recognition and association can be implemented in a computer program. However, it should be pointed out that the level of play of CHUMP is rather low. This is because it plays chess by pure pattern recognition, without being able to look ahead, and it is well established that look-ahead abilities are important in playing chess and other board games at a high level [20].

5.2   Expectations

Expectations are important in guiding what we look at and how effectively we can recognise what we see. Expected objects are recognised with greater accuracy than unexpected objects, particularly in noisy domains [21]. Expectations may also relate to complex collections of objects, or *schemata*. Perceptual classification of objects within a familiar schema can be quicker than when the objects are not in the schema. For instance, Biederman [22] describes an experiment in which participants took longer to identify a fire-hydrant when positioned above street level than when at its expected position. Finally, noisy or ambiguous scenes may be reconstructed, if the visible elements are constrained to fit a compatible schema; Lindsay and Norman [23] describe such an experiment with words composed of distorted or ambiguous letters.

The main lesson from the results described above is that perception is not a simple flow of information from scene to memory, but instead perception is an active process, with the attention forcing a shift of fixation point to different parts of the scene based partly on what is observed and partly on what is expected (or anticipated) to be present. CHREST has been used to construct a model of this process, and demonstrate the qualitative results described above [24]. The key factor in supporting this process was for CHREST to support links between visual information, representing the scene being analysed, and verbal

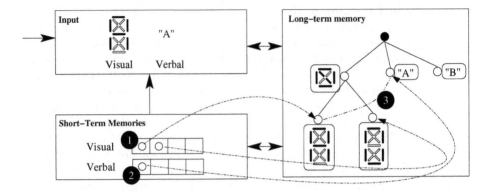

Fig. 4. Learning to associate information across two modalities: (1) visual pattern is sorted through LTM, and a pointer to the node retrieved is placed into visual STM; (2) verbal pattern is sorted through LTM, and a pointer to the node retrieved is placed into verbal STM; and (3) a naming link is formed between the two nodes at the top of the STMs. (Taken from [24].)

information, representing the interpretation of the scene. The links between these two input modalities are formed when the model is presented with visual and verbal information simultaneously, as illustrated in Fig. 4.

In the situation depicted, the model is presented with a visual stimulus and a verbal label. Both the stimuli are sorted through the long-term memory, and pointers to the retrieved nodes are placed in their respective short-term memory. A *naming link* is formed between the two nodes at the top of the two short-term memories. This naming link can be used in future to retrieve the verbal label from a visual stimuli. By a similar process, sequences of letters and words can be acquired and stored in the network.

The attention mechanism can use this stored information in various ways. First, verbal priming (e.g. being told to find particular words) can be used to highlight specific visual chunks, which the attention mechanism will then attempt to locate. Second, when part of a chunk is located, the visual or verbal information can be used to identify schemata which can guide the attention mechanism. For instance, verbally learning a sequence of words can be used to trigger recognition of the visual representations of those same words.

6   Discussion

In this paper, we have described CHREST and shown how it provides an account of the links between perception, attention, learning, memory and action. Thus, CHREST addresses several key issues in the study of attention, and in particular how it links to other aspects of human cognition. One of the differences of CHREST with respect to other theories of attention is the level at which it works; CHREST is a *symbolic* system, meaning that information is stored at a 'meaningful' level. The basic information used by CHREST in the chess experiments is the *piece-on-square*, which encodes that a piece is located at a particular position of the board. This perspective distinguishes our work from that by other authors on perception, such as Tsotsos [25] or Wolfe [26]. CHREST's perspective is psychological, but related to high-level conceptual processing, and is not concerned with neurological or other low-level processes.

The key contribution of this paper is to show how a symbolic treatment of attention can affect our understanding of attention in high-level areas of human cognition, and we now review some of these areas.

Perception is cognition. This was the central focus of the studies reported in de Groot and Gobet's book [5]. This role has sometimes been overlooked in recent attempts to provide general architectures of cognition. For example, in both Soar [27] and ACT-R [28], perception is considered, but its role is not as central as in CHREST. As we have seen, in CHREST the central theoretical construct is that of a perceptual chunk, and the learning of chunks directly depends on perceptual mechanisms.

Attention directs learning, and that which has been learned directs attention. This relates to the previous point. Expertise in a domain develops because the use of fairly weak heuristics leads to the acquisition of more perceptual knowledge. These heuristics can be either general ("look at a part of the scene you don't know anything about") or domain-specific ("verify whether your Queen is attacked"). With sufficient experience, this perceptual knowledge enables the fluid and rapid behaviour that is characteristic of experts' intuitive decision making.

Action is closely linked to perception. The entire CHREST architecture rests on the assumption that many actions are elicited by perceptual patterns. This is particularly the case with expert behaviour, where the link between action and perception is made automatic.

Attention is directed by a combination of top-down and bottom-up processes. The CHREST simulations show how novices rely more on weak heuristics to direct their attention, while the masters rely more on the structure of the discrimination network to suggest the next fixation position. Note that some of the (top-down) heuristics used by the masters may also be useful as information-seeking devices. Top-down heuristics are also important to make sure that global goals are heeded.

Context matters in focusing attention. In CHREST, this is readily captured by the fact that the fixations directed by the discrimination network are sensitive to even fine details in the context. Indeed, different positions lead to different eye movement patterns. Just like human masters, CHREST does not use stereotyped sequences of eye movements, but adapts these as a function of the environment.

Attention is crucial at many levels. While this paper has focused on the role of attention in the first seconds of presentation of a new complex stimulus, it should be pointed out that, in a full model, attention would be essential at several levels. For example, when chess players examine a position in order to choose the best move, attention would be selective, not only in selecting the next fixations, but also in deciding which moves should be selected for further consideration, and also in deciding what part of the information currently gathered about the position is worth the cost of storing in long-term memory. In both cases, selectivity – and not just random guess – is made possible by knowledge acquired over years of dedicated practice [15].

Why is attention limited? And why is the capacity of short-term memory limited? From an evolutionary standpoint, limits in attention and short-term memory capacity make sense; when facing a danger, organisms that immediately identify the presence of the danger are more likely to survive (and reproduce) than organisms that have to process the many stimuli they focus on, and sift through masses of information in their short-term memory. Indeed, Simon [29] argued that selective attention is a key feature of the broader notion of bounded rationality – the assumption that humans make decisions in science, business and everyday life with only a small amount of search. Thus, selective attention enables rapid decisions that may not be optimal, but that are good enough.

While CHREST is a computational architecture of cognition that captures many of the aspects of attention as studied in cognitive psychology, it was not developed as a model of attention only. As we have seen earlier, it also captures many aspects of (high-level) perception, learning, memory and decision-making. CHREST may also have interesting things to say about machine learning and robot perception. For example, chunking turns out to be a robust statistical learning mechanism, and we have provided several examples in this paper of how chunking is linked to attention.

The way attention is used in CHREST can also shed light on questions central to the development of autonomous intelligent systems. The frame problem is such a question. How can a system notice the relevant changes in the environment, while ignoring the irrelevant changes? Due to the various limitations that characterise CHREST, it actually does not face the frame problem. Together, CHREST's attentional restrictions implemented by a limited capacity memory enable it to pay attention just to a few features of the environment. With increasing expertise in a domain, the perception-learning-perception cycle we have discussed above leads to increasingly rapid and adequate decisions, whilst not necessarily increasing the attention span of the system.

Evolutionary considerations highlight two factors important in modulating attention, which we have started incorporating into CHREST: emotions and motivations. In a classic paper, Simon [30] argued that, with systems characterised by serial organisation and control hierarchy, motivation refers to what is controlling attention at a specific time; in particular, given that these systems have multiple goals, motivation controls how attention is focused on a specific goal. Furthermore, it is necessary to have a provision for interrupt mechanisms; Simon proposes that at least two sources inform these mechanisms: first, drives (for example, hunger), and, second, the information gained by EPAM's process of noticing. In particular, emotional tags might be added to perceptual chunks during the learning process. Later on, the emotions associated with these chunks, in particular when they are negative, such as fear, may direct attention to specific aspects of the environment. For example, a chess master might have suffered a painful loss in a given type of position. In future games with the same type of position, her attention is likely to be modulated by the knowledge of this previous game, and the emotions associated with it – for better or for worse.

7 Conclusion

In this paper, we have described how the CHREST cognitive architecture explains, simulates and employs attention mechanisms. Two aspects of CHREST's implementation are central to explaining the attention mechanism in humans: the first is the tight cycle of perception-learning-perception; the second is the use of discrete *chunks* of information, both in long-term and short-term memory. The perception-learning-perception cycle guides the attention mechanism through a set of heuristics, which select a new fixation point by combining bottom-up and top-down information. The discrete nature of chunks enables a limited short-term memory to refer to a far larger store of information, and also supports the use of cross-modal or image-action associations. As CHREST's problem-solving and action abilities are further extended, we anticipate greater insights will emerge about the link between an expert's selective attention and their ability to produce rapid and skillful responses.

Acknowledgements

We thank the reviewers for their comments on a previous version of this paper.

References

1. de Groot, A.D.: Thought and Choice in Chess. The Hague, Mouton (1946/1978)
2. Simon, H.A., Gilmartin, K.J.: A simulation of memory for chess positions. Cognitive Psychology 5, 29–46 (1973)
3. Gobet, F., Lane, P.C.R., Croker, S.J., Cheng, P.C.H., Jones, G., Oliver, I., Pine, J.M.: Chunking mechanisms in human learning. Trends in Cognitive Sciences 5, 236–243 (2001)
4. Feigenbaum, E.A., Simon, H.A.: EPAM-like models of recognition and learning. Cognitive Science 8, 305–336 (1984)
5. de Groot, A.D., Gobet, F.: Perception and Memory in Chess: Heuristics of the Professional Eye. Van Gorcum, Assen (1996)
6. Gobet, F., Simon, H.A.: Templates in chess memory: A mechanism for recalling several boards. Cognitive Psychology 31, 1–40 (1996)
7. Gobet, F.: Expert memory: A comparison of four theories. Cognition 66, 115–152 (1998)
8. Gobet, F., Simon, H.A.: Five seconds or sixty? Presentation time in expert memory. Cognitive Science 24, 651–682 (2000)
9. Freudenthal, D., Pine, J.M., Gobet, F.: Modelling the development of children's use of optional infinitives in English and Dutch using MOSAIC. Cognitive Science 30, 277–310 (2006)

10. Minton, S.: Quantitative results concerning the utility of explanation-based learning. Artificial Intelligence 42, 363–391 (1990)
11. Lane, P.C.R., Cheng, P.C.H., Gobet, F.: Learning perceptual schemas to avoid the utility problem. In: Bramer, M., Macintosh, A., Coenen, F. (eds.) Research and Development in Intelligent Systems XVI: Proceedings of ES 1999, the Nineteenth SGES International Conference on Knowledge-Based Systems and Applied Artificial Intelligence, Cambridge, UK, pp. 72–82. Springer, Heidelberg (1999)
12. Gobet, F.: Discrimination nets, production systems and semantic networks: Elements of a unified framework. In: Proceedings of the Second International Conference of the Learning Sciences, pp. 398–403. Northwestern University, Evanston (1996)
13. Burl, M.C., Asker, L., Smyth, P., Fayyad, U., Perona, P., Crumpler, L., Aubele, J.: Learning to recognize volcanoes on Venus. Machine Learning 30, 165–194 (1998)
14. Talavera, L.: Feature selection as a preprocessing step for hierarchical clustering. In: Proceedings of the International Conference on Machine Learning (1999)
15. Chase, W.G., Simon, H.A.: Perception in chess. Cognitive Psychology 4, 55–81 (1973)
16. Gobet, F., Clarkson, G.: Chunks in expert memory: Evidence for the magical number four... or is it two? Memory 12, 732–747 (2004)
17. Gobet, F.: Is experts knowledge modular? In: Proceedings of the Twenty-Third Annual Meeting of the Cognitive Science Society, pp. 336–341. Lawrence Erlbaum, Mahwah (2001)
18. Campitelli, G., Gobet, F.: Adaptive expert decision making: Skilled chessplayers search more and deeper. Journal of the International Computer Games Association (2004)
19. Gobet, F., Jansen, P.: Towards a chess program based on a model of human memory. In: van den Herik, H.J., Herschberg, I.S., Uiterwijk, J.W. (eds.) Advances in Computer Chess 7, pp. 35–60. University of Limbourg Press, Maastricht (1994)
20. Gobet, F., de Voogt, A., Retschitzki, J.: Moves in Mind: The Psychology of Board Games. Psychology Press, Hove (2004)
21. Neisser, U.: Cognitive Psychology. Appleton-Century-Crofts, New York (1966)
22. Biederman, I.: On the semantics of a glance at a scene. In: Kubovy, M., Pomerantz, J.R. (eds.) Perceptual Organization, pp. 213–254. Lawrence Erlbaum, Hillsdale (1981)
23. Lindsay, P., Norman, D.: Human Information Processing. Academic Press, New York (1972)
24. Lane, P.C.R., Sykes, A.K., Gobet, F.: Combining low-level perception with expectations in CHREST. In: Schmalhofer, F., Young, R.M., Katz, G. (eds.) Proceedings of EuroCogsci., pp. 205–210. Lawrence Erlbaum, Mahwah (2003)
25. Tsotsos, J.K., Culhane, S.M., Wai, W.Y.K., Lai, Y., Davis, N., Nuflo, F.: Modeling visual attention via selective tuning. Artificial Intelligence 78, 507–545 (1995)
26. Wolfe, J.M.: Guided search 2.0: A revised model of visual search. Psychonomic Bulletin and Review 1, 202–238 (1994)
27. Newell, A.: Unified Theories of Cognition. Harvard University Press, Cambridge (1990)
28. Anderson, J.R., Bothell, D., Byrne, M.D., Douglass, S., Lebière, C., Qin, Y.L.: An integrated theory of the mind. Psychological Review 111(4), 1036–1060 (2004)
29. Simon, H.A.: A behavioral model of rational choice. The Quarterly Journal of Economics 69, 99–118 (1955)
30. Simon, H.A.: Motivational and emotional controls of cognition. Psychological Review 74, 29–39 (1967)

Modeling the Interactions of Bottom-Up and Top-Down Guidance in Visual Attention

David Henderickx, Kathleen Maetens, Thomas Geerinck, and Eric Soetens

University of Brussels, Faculty of Psychology and Educational Sciences,
University of Brussels, Faculty of Electronics and Informatics,
Pleinlaan 2, 1050 Brussels, Belgium
http://www.vub.ac.be/PE

Abstract. We propose a framework of visual attention grounded on earlier attentional and perceptual models to serve as a basis for the development of computational vision systems. The framework is build on generally established knowledge about the neural basis of human attention, models developed by Briand [1][2], Itti and Koch [3][4], and Wolfe [5][6] and our own results that call for adaptations of the existing models [7][8]. In this paper we concentrate on the interaction of bottom-up and top-down processes to understand the mechanisms underlying exogenous and endogenous attention. Two series of studies are reported to support the proposed adaptation of the earlier models. First, we claim that the visual feature binding is a common process for exogenous and endogenous attention [7]. Secondly, we demonstrate the ability to preset the bottom-up feature maps by demonstrating the phenomenon of Inhibition of Return (IOR) with endogenous cueing, suggesting that IOR affects processing before focusing attention [8].

Keywords: visual attention, endogenous attention, exogenous attention, feature integration, inhibition of return.

1 Introduction

An essential research topic in the area of 'Computational Vision' is the development of an efficiently working attention system. Several models of attention in computer vision have been suggested in earlier studies, but only few of these models have incorporated mechanisms for both bottom-up, image-activated processes, and top-down, task-depending processes. In the present paper we make suggestions for alterations to these models, specifically related to the interaction of bottom-up and top-down processes in early vision.

Research over the last three decades has demonstrated that selection of incoming information in humans is achieved by the allocation of spatial attention, facilitating the processing of selected objects without shifting gaze direction [9][10][11]. Making attentional shifts allows us to combine chunks of sensory information resulting from localized visual analysis problems, to build a mental representation of our visual world. The allocation of spatial attention appears

L. Paletta and J.K. Tsotsos (Eds.): WAPCV 2008, LNAI 5395, pp. 197–211, 2009.

to be jointly determined by task-dependent (endogenous) and image-based (exogenous) factors [12].

1.1   Exogenous and Endogenous Attention: Bottom-Up Versus Top-Down

First, spatial attention can be voluntarily directed. We actively and deliberately select and focus on what we believe is important, usually aimed at achieving an immediate visual goal (goal-driven). This is called *endogenous attention* and is considered *top-down*, because higher brain regions are involved at the outset [9][13][14]. According to Posner and Snyder [15] and Jonides [16], top-down orienting is resource-limited, and is affected by expectancies and concurrent memory load.

A second important mode of operation is largely unconscious and driven by external objects (or their specific attributes) that stand out from their environment. These conspicuous stimuli attract attention automatically (*attentional capture* [16][17]) and independently of intentions [14][18], even if the stimulus is irrelevant for the ongoing task. The ability to allocate attention rapidly to salient objects is of particular behavioral importance, because it constitutes a powerful alerting system, allowing the organism to detect quickly a possible prey, mates, or predators in our environment. This mode of attentional control is known as *exogenous attention* and is considered to function primarily in a *bottom-up manner*, because principally it operates stimulus-driven. Bottom-up orienting (as a process) is resource-free, cannot be suppressed, is unaffected by subject's expectancies or by concurrent memory load and does not require conscious awareness[1] [15][16].

2   Disclosing the Mechanisms Underlying Exogenous and Endogenous Attention

In a first series of studies, we focused on the work and model of Briand and Klein concerning the role of spatial attention in *visual feature binding* [1][2][19]. Treisman's feature integration theory (FIT) was an attempt to define one of the purposes of focused attention [20]. Treisman and colleagues found that certain elementary object features such as color or form, could be detected massively parallel over the entire visual field, while conjunctions of these features could not. They hypothesized that for detecting elementary visual features pre-attentive processing is sufficient, but that focused attention is necessary for binding the elementary visual features of an object into a unitary percept that enables proper identification and detection of the object.

---

[1] Note that in section 3.1, we will demonstrate that it is possible to realize top-down presetting of the low-level input maps, which are the source of bottom-up orienting. Accordingly, an upcoming stimulus-driven orienting process has already been affected by top-down influences before it has started.

Briand and Klein wondered whether the focused attention needed in feature binding ('Treisman's *glue*') is the same as Posner's attentional spotlight that enhances the efficiency of the detection of events within its *beam* [1][2][9]. Therefore, Briand and Klein combined a variant of Treisman's feature/conjunction task with Posner's (cost-benefit) orienting paradigm, making use of both endogenous and exogenous cueing conditions. Briand et al. found that spatial cueing effects were larger for conjunction targets than for feature targets, but only in the exogenous condition. In the endogenous cueing condition, no such interaction between cue validity and search type (feature-conjunction) was present. With endogenous cueing, the cost-benefit pattern was equivalent for feature and conjunction search.

According to Briand et al., the interaction between cue validity and search type, observed exclusively in exogenous cueing, suggests that exogenous attention affects the spatial attentional system that performs the binding function attributed to spatial attention by Treisman's FIT [20]. Because the endogenous cueing condition does not show this interaction, endogenous attention must be based on a different attentional mechanism, which operates independent of feature binding.

Briand proposes an attentional framework grounded on the idea of the existence of separable attentional subsystems: exogenous attention being controlled by the posterior attention system and endogenous attention being controlled by the anterior attention system [2]. This idea was promoted by Posner and Petersen and involves the assumption that each system controls different processing aspects of attentional orienting [21].

Note that in this framework (see Figure 1), the terms exogenous and endogenous respectively correspond with the posterior and the anterior attention system and that only exogenously controlled spatial attention directly affects feature

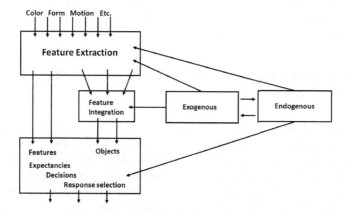

Fig. 1. Schematic model illustrating the processing stages that may be influenced by exogenously controlled (posterior) and endogenously controlled (anterior) attention systems (Taken from Briand [2])

integration. The endogenous system acts upon several stages in perceptual processing and is able to control and modulate the exogenous attention system.

2.1 Challenging Briand's Study: Global Pop-Out

In our studies however, we challenge Briand's conclusions, based on results of variations of Briand and Klein's experimental design [7]. Briand presented two colored letters at one side of the fixation point. Such a target differs significantly from its surrounding, and will capture attention in a bottom-up way. The induction of such a *global pop-out* could have altered cueing effects in Briand's results, because pop-out always directs attention automatically to the valid location of the upcoming target. Since endogenous cues do not direct attention automatically, they can easily be ignored to the advantage of the global pop-out. As a result, the endogenous cueing effect will weaken and thus also the interaction between cue validity and search condition will drop. In an exogenous condition, the cue is processed automatically and cannot be ignored. Consequently, global pop-out cannot exert such a strong influence on the effect of cueing, and accordingly the cue validity - search type interaction will be less violated. In sum, a discrepancy in the effect of a global pop-out between both cueing conditions could seriously misrepresent a true comparison of both attentional modes.

We first simulated Briand's design, i.c. task-relevant stimuli were presented on a single side of fixation. In a second experiment, we adjusted Briand's design presenting task-relevant stimuli on both sides of fixation. In Experiment 1, we replicated Briand's results, whereas in Experiment 2 with 2-sided stimulus presentation, results showed the presence of a target type x validity interaction for both endogenous and exogenous cueing. Thus, when in the endogenous cueing condition the reaction time pattern is not distorted by the automatic attentional capture by means of a single sided task-relevant stimuli presentation, a target type x validity interaction is found. This suggests that although exogenous and endogenous attention have different origins, they influence stimulus binding in a similar way.

3 Inhibition of Return: High-Level or Low-Level Mechanism?

A second adaptation of the early attentional models concerns the phenomenon of *IOR*. This is an inhibitory mechanism of attention, which is thought to promote exploration of previously unattended objects in the scene during visual search or foraging by preventing attention from returning to already-attended objects or locations. Some early models of artificial vision situate IOR at the level of focused attention [3][4]. The assumption is that attention is drawn to the most salient feature in a visual display in a bottom-up manner. After attention has been focused there by a 'winner-takes-all' mechanism, the item is identified as being a target object or not. If this most salient 'object' is not a target, a feedback loop is assumed, inhibiting the object location in the saliency map, and the next most salient object is selected. As the inhibitory process is rather object-based, since a location can be inhibited only after the object-identification process has finished, high-level visual processing is essential.

Although such a high-level *object-based* IOR mechanism may exist, it does not correspond to the IOR phenomenon observed in most attention research. In these IOR studies, the subject is precued on the 'location' of a subsequently appearing target, which participants have to detect or localize as quickly as possible. Typically both the validity of the location cue (valid vs. invalid) and the time between precue and target onset (cue target interval, CTI) are manipulated. First, attention is captured by the onset of a peripheral location cue which predicts with a certain validity the location of the soon appearing target, e.g. a brief illumination of the frame around one of the possible target locations, that automatically captures attention due to its sudden onset. When the CTI is larger than approximately 250ms, individuals are slower to respond to a target at the location that has been previously cued (valid cue) compared to a target at a location that has not been cued (invalid cue). This IOR reaction-time pattern is the reverse of a normal cueing-effect, in which benefits are observed with validly cued targets and costs with invalidly cued targets. This pattern of results seems to indicate that after attention is reoriented back to the center of the screen, subjects seem to give attentional preference to locations where attention has not been guided before. Since the target has not yet appeared at the time attention is reoriented to the center of the screen, no high level feature binding and target identification processes have been activated yet. Accordingly, if inhibition of subsequent processing at the cued location is observed, the inhibition process is situated before feature binding. This *location-based* inhibition of previously attended locations is mostly found with exogenous cueing, but not with endogenous orienting of attention. We conducted a series of 4 experiments, in which we manipulated low-level visual processing by means of endogenous cues [8]. Herewith, we tried to demonstrate that IOR is triggered only when the low-level visual routes are being activated.

3.1   Low-Level Location Inhibition Affects Processing before Focusing Attention: Presetting the Bottom-Up Feature Maps

Sensory information is distributed to several distinct retinotopically organized visual feature maps (see [22][23] for neurological support). To locate the most conspicuous feature, features are mutually compared for each feature dimension (color, orientation, motion). Accordingly, a conspicuity map for each of these dimensions is constructed. Eventually the most conspicuous feature, derived from the combination of conspicuity maps, will serve as bottom-up guidance for the attentional system, and is responsible for attentional capture.

After attention is drawn to the most conspicuous feature-location of the visual display in a bottom-up manner, that particular location is inhibited and/or other locations in the visual display are privileged for being attended to. With endogenous orienting our intentions rather than the conspicuity of features, guide attention. Consequently, usually no low-level inhibition process is triggered. This would explain why in general no IOR occurs under endogenous cueing conditions. We assume that if a location-bound IOR is indeed situated at low-level visual processing and if we can guide attention intentionally (top-down) by modulating the low-level input maps (visual buffer), it should be possible to find location-bound IOR also with endogenously triggered orienting.

In order to realize top-down presetting of the low-level input maps (also see: Attentional Set [24][25][26][27][28]; Guided Search [5][6]), we set up two experiments with two possible target locations, left and right from fixation. Before the target was presented, there were two cues: a primary color cue, that informed the subjects of the use of the second cue, and a location cue that shortly marked the two possible target locations with a different color [8]. Top-down information was provided by means of the color cue presented prior the location cue. That is, the color cue informed the subject of which color of the location cue they had to pay attention to, before they had to detect the target ('endogenous' cueing). Since with the location cue both possible target locations were simultaneously marked by a color flash, the cue did not attract attention exogenously to one particular location. When the target appeared at the location where the second cue color was identical to the primary cue color, the experimental trial was considered as a valid cued trial. When the target appeared at the other location, the trial was considered invalid. Subjects were never informed about the cue validity, which was 50 percent. During the experiment subjects were online checked whether they carried out the cueing instruction properly. In the first experiment, the primary cue was presented once before a large number of trials (*blocked*). In the second experiment, the primary cue color was *randomized* and presented before each trial. In both experiments, CTI was manipulated. When the CTI was short, subjects reacted faster to a target at the cued location compared to the uncued location. However, when CTI was longer, we found the expected IOR effect, that is, faster reaction times for targets at the uncued location compared to the cued location. Thus, we observed the assumed IOR under endogenous cueing for the long CTI conditions and a normal cue facilitation-effect after short CTI's.

As a control, another two similar endogenously cued detection tasks were set up, but without presenting the color information prior to the location cue. In this way, no presetting of the low-level input map was possible. In a first control experiment, the colored peripheral flashes and the central endogenous cue were presented simultaneously. In a second experiment, the endogenous cue was presented after the location cue. In these control conditions, subjects did no longer have the time to preset the low-level input maps. As expected, we did not find any IOR, but a normal facilitation cueing effect was found for all CTI.

Although in general, no IOR is observed with endogenous cueing, we were able to demonstrate the phenomenon by rerouting top-down endogenous attention to the low-level input maps. Hereby, we conclude that location-IOR is situated at the low-level visual processes and needs to be distinguished with the possible coexistence of a higher order IOR [3][4].

4   Consequences for the Existing Attention Frameworks

Considering our findings in relation to the results of Briand and our conclusions regarding a low-level IOR, revising the earlier models is inevitable for understanding the attentional network and essential for developing a *humanoid attention system*. In the following chapter, we propose an adapted two component framework of visual attention, grounded on (a) the attentional or perceptual models of Briand and Klein [1][2][19], Itti and Koch [3][4], and Wolfe [5][5], (b) the conclusions based on our own studies and their impact on the original models [7][8], as well as (c) reasonably well-established facts about the neural basis of attention.

4.1   Assembling a Two Component Attentional Framework

As discussed earlier, visual stimuli have two ways of attracting attention and consequently penetrating to higher levels of awareness: being volitionally brought into the attentional focus (top-down), or by means of automatic attentional capture (bottom-up). However, one has to keep in mind that in everyday life both attentional systems are continuously interacting. In Figure 2 we developed a tentative model for visual attention, based on the earlier models and complemented with adaptations based on our research results.

4.2   Neuroscientific Grounds

Much evidence has been accumulated in favor of a two-component framework for the control of attentional orienting and focusing within a visual scene. Neurophysiological studies with monkeys and data from brain-injured patients have indicated that the *posterior parietal cortex* [29][30], lateral pulvinar nucleus of the thalamus [31][32], and superior colliculus [21][33] are involved in attentional orienting and focusing. On the other hand, the *anterior cingulated cortex* (a) appears to play a special role in target detection and selection of task relevant stimuli [34][35], (b) has major connections to the amygdala, which plays a critical role in emotion [36], (c) to the *posterior parietal cortex* and possibly the pulvinar nucleus, (d) and even to dorsolateral prefrontal cortex and parahippocampal cortex, which are involved in short-term visual-spatial and long-term object memory, respectively [37]. The anatomy and neurophysiology of the anterior cingulated suggests that this system provides top-down assistance, out of its involvement with several other cognitive functions, to the posterior attention system, which operates on the basis of simple stimulus features and automated routines.

4.3   Attentional Selection: Bottom-Up Guidance of Attention

In bottom-up guidance of attention, the process starts with and is guided by external stimuli in the visual field. In a first phase, as in most models of visual attention, visual features (horizontal, slanted, red, blue,) are extracted pre-attentively and in parallel from the retinal image (see Figure 2). The sensory information is distributed to several retinotopically organized *visual feature*

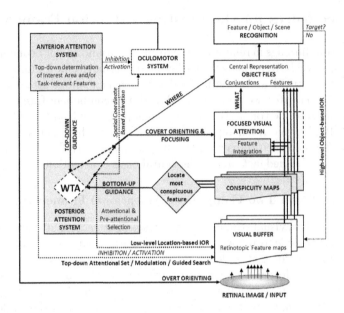

Fig. 2. The adapted two-component, winner-take-all (WTA) framework of spatial visual attention grounded on the models of Briand and Klein [1][2][19], Itti and Koch [3][4], and Wolfe [5][6], and based on our behavioral studies [7][8]

maps. Note already, that we do not imply that feature extraction is purely feed forward, as top-down influences from information, gathered about the retinal image and/or the target searched for, can possibly preset the feature maps (Guided Search, [5][6]; Attentional Set [24][25][26][27][28]; and [8]). The processing in this stage is capacity unlimited and attention is diverged.

In a next phase, the retinotopic visual feature maps are combined per feature dimension (e.g. color, orientation, motion) to detect conspicuous features. The localization of conspicuous features is implemented by *center-surround* difference mechanisms for each feature dimension (see also [3][4]). Activities in the resulting *conspicuity maps* are mutually compared to locate the most conspicuous feature over all feature dimensions. Eventually the most conspicuous feature, derived from the combination of conspicuity maps, will serve as bottom-up guidance for the posterior attention system, which will determine the target of attentional orienting and focusing. As a consequence, unique anomalies within a feature dimension will automatically attract our attention[2]. This phenomenon is named "attentional capture" or also "the pop-out effect". It is also in this phase that

---

[2] Note that conspicuity maps can have different weights by nature (e.g. the color map has a higher weight than the orientation map) and additionally can be pre-attentionally dynamically adjusted by top-down attentional biases or prior experience through inhibition or activation of certain feature maps (e.g. when we are looking for red apples at the greengrocer's, our susceptibility for all what is red will increase and/or our sensitivity for other features will be inhibited).

unbalanced presentation of stimuli can elicit an unwanted attentional capture (or bottom-up guidance) in cueing experiments. When comparing endogenous and exogenous cueing effects, this global pop-out due to the presentation of salient stimuli, will primarily interact in conditions intended to measure endogenously guided attention. This influence was clearly demonstrated in our first series of experiments.

4.4 Attentional Selection: Top-Down Guidance of Attention

Attention can also be directed by top-down information from intentions created by prior knowledge (strategy, instructions, experience). In this case, the posterior attentional system is guided by the higher-level anterior system that tries to determine the focus of attention. Accordingly, the posterior attention system is common for both exogenous and endogenous attention in the model. This common system is responsible for the actual orienting of the attentional focus to visually informative locations within our visual field [21][33], and affects feature integration. This arrangement is grounded on the results of our experiments in this study: we demonstrated that although exogenous and endogenous attention have different origins, they influence stimulus binding in the same way. Note, that this reasoning implies that we no longer treat endogenous/exogenous and anterior/posterior, respectively, as analogous terms.

4.5 Interacting Attentional Routines: Winner-Take-All

In the posterior attentional system, the location that will be attended to is determined by the result of the *winner-take-all* (WTA) interaction between the 'endogenous' top-down guidance and the 'exogenous' bottom-up guidance routine. On the one hand, internal (top-down) information from other cognitive functions acts upon the attentional orienting system, i.e. the posterior attention system. On the other hand, externally received low level visual information tries to modulate the orienting of attention through a bottom-up mechanism of conspicuousness. A winner-take-all arbitration between both attention modes is proposed, since either the most conspicuous object derived from the exogenous routine or the most dominant object (or area) arisen from the endogenous routine, is attentionally selected for further processing. Note however, that this does not exclude the possibility of a serial combination of both attentional routines: attention may be oriented endogenously to an area after which the most conspicuous object within that area is selected.

By strategically activating and/or inhibiting retinotopic feature maps (i.c. top-down attentional set), the bottom-up visual processing routine can also be modulated by endogenous factors. For instance, when you know that what you are looking for is blue, the visual system will favor stimuli that activate the blue feature map for further processing. This will also favor all what is blue in the battle for most conspicuous feature. Yantis and Jonides [38] found that sudden onsets do not necessarily attract attention automatically (exogenous/bottom-up), due to endogenous control modulation. Our studies with endogenous cues further support the claim that feature maps can be strategically preset. Knowing in advance which color is the relevant cue, creates the opportunity to preset that particular feature map.

4.6   Inhibition of Return (IOR)

IOR is a process that directs our attention to locations that have not previously been attended to, rather than to the previously attended locations. As we discussed earlier, this inhibitory mechanism cannot be located at the level of stimulus identification and is probably situated before the process of feature binding. IOR is assumed to be an automatic process, guided by the posterior attentional network. For these reasons we expect the mechanism to operate at the bottom-up attentional route after the WTA-interaction (see Figure 2). At this stage, the battle for attention has been won by the most conspicuous feature and the attention is being focused to its location. After a delay, attention will be oriented back to a neutral position (fixation) by the posterior attention system. At that time the posterior system feeds back to the retinotopic input maps, in order to inhibit the previously attended (most conspicuous) location.

The localization of IOR in the bottom-up information stream is further supported by our series of experiments with endogenous cues. Giving subjects preknowledge about the color to be looked for, resulted in a regular cost-benefit pattern for short CTI's, and an IOR (i.c. a reversed cost-benefit pattern) for longer CTI's, just as it does for exogenous cues. In general IOR is not found in endogenous cueing tasks without top-down attentional set. This shows that IOR is located before the process of focused attention, since focused attention is a common element in the exogenous and endogenous attentional route.

Since we do not want to exclude the possibility of a high-level object-based IOR process in the adapted attentional framework, a top-down object location inhibition path is suggested (see Figure 2). After an object has been identified as being a non-target, the object location is retrieved from the object file and subsequently top-down (voluntarily) inhibited in the retinotopic input maps.

4.7   The Oculomotor System

As most attentional theories assume a strong relationship between visual attention and saccades, an overt attention module is suggested. An oculomotor system is driving the saccadic eye movements and is accordingly determining the part of our visual environment that feeds the retinal image. In the current framework, this oculomotor system can be activated through top-down processes, through bottom-up processes, as well as through the combination of bottom-up and top-down processes.

First, our eyes can be intentionally (top-down) guided without visual stimuli in the environment. For instance, it is possible to intentionally direct our eyes to the left or to the right without seeing anything (e.g. in a dark room).

Second, our eyes can be guided automatically to the area of a conspicuous object or feature that captured attention through bottom-up attentional guidance.

Accordingly, the goal of the saccade is determined by the spatial coordinates of the conspicuous object. However, since we are able to inhibit making eye movements, top-down processes are also able to inhibit the oculomotor system from directing a saccadic eye movement to the most conspicuous object area.

A final oculomotor trigger, and probably the most common trigger, is the combination of both top-down and bottom-up guidance (cf. Guided Search [5][6]). Clark suggested that voluntary eye movements occur through high-level modulation of the low-level substrate that underlies the reflexive bottom-up guided saccadic eye movements [39]. In the framework, this kind of modulation was already made possible through the top-down attentional set route, modulating the low-level retinotopic input maps. Support for this idea is found in the observation of IOR in endogenous cueing experiments when the endogenous cues trigger the oculomotor system [40][41]. In general, no IOR is found with endogenous cueing. However with top-down modulation of the low-level input maps, the IOR phenomenon could also be observed after endogenous cueing [8].

4.8   Focusing Visual Attention: Feature Binding and Further Visual Computation

As explained before, focused visual attention is required to ensure the correct integration of features (feature binding) to specify objects. When object features are processed, the integrated and complex object properties are stored in *object files*, together with their spatial location. The central representation of the retinal image is constituted of these serially accumulated object files, which are approachable from our memory to realize object and scene recognition. Note that in the model, the object properties (*"what"*) and their location (*"where"*) originate from *focused vision* and the 'posterior attention system' respectively. The dual route idea is taken from Ungerleider and Mishkin [42]. Their research indicated that the visual cortex is organized into two distinct pathways both originating in the primary visual cortex. The ventral stream which reaches the inferotemporal cortex is involved in the identification of objects, whereas the dorsal stream, which projects into the parietal cortex, is engaged in the visual spatial localization of objects (for an alternative view, see [43]). Single-cell recording studies in monkeys have found cells in inferior temporal cortex to fire in response to shape or color, while cells in posterior parietal cortex fire in response to location, size or motion of an object [44]. From introspection, as well as from several psychophysical studies, we know that we do not only notice the world within our focus of attention. We are able to make simple judgements about non-attended visual elements, although those judgements are limited and less accurate than those made in the presence of attention [4][20][45]. In the model this is represented by direct links between the visual buffer and the object files, where there is no connection with the process of feature integration (see Figure 2). Notice that this implies that judgements only can be made on the basis of raw unbound features. Even when a feature triggers an exogenous orientation, no attentional resources are required to make a judgement about this single feature, in contrast with conjunction judgements, where feature integration demands additional attentional resources. Since attentional orienting and focusing is shortcut, simple feature judgements will be faster than conjunction judgements. This double pathway provides us with an explanation for the interaction between target type (feature and conjunction) and cue condition (valid or invalid), reported in the exogenous condition of Briand and Klein [1][2] and in our endogenous and exogenous studies, where we avoided the occurrence of a global pop-out.

5   General Conclusion

The simulation of a visual attention network in computational vision systems is a complex issue, and has resulted in the development of several models, each developed with their own specific accent or target application. The present suggested adaptations have been based on earlier models developed by Wolfe [5][6], Itti and Koch [3][4], and Clark [39]. More specifically, we have concentrated on the interaction between bottom-up and top-down mechanisms of information processing and the suggested ideas in this respect expressed in the model of Briand [2] and Klein [19]. The adaptations that we suggest are based on two series of experiments of which we have explained the general outlines and results.

In a first series of experiments we demonstrated that the integration of stimulus features can be controlled by both exogenous and endogenous attention cues, implicating that the anterior attentional system controls the posterior attentional system, and in this way indirectly influences feature binding. The implications for modeling are that exogenous and endogenous attention have to be explained in terms of interactions between an anterior and a posterior attentional network and how these networks are related to the bottom-up and top-down information processing streams. In a second series of experiments we demonstrated that IOR, which has mostly been linked to exogenous cueing, can also be elicited with endogenous cues. These results can be explained by assuming that (a) low-level input feature maps can be preset on the basis of top-down influences if there is sufficient time for presetting before the bottom-up process starts, and (b) that IOR is also a relatively early attentional mechanism, situated at the level of conspicuity maps, but is nevertheless also possible for endogenous cues because of the presetting of feature maps.

Finally, an effort was made to incorporate an oculomotor system that is related to the covert attention system, and is accounting for overt attentional phenomena. The overt attentional system determines the selection of the visual environment that feeds the retinal image and provides the visual system with external visual data.

The suggested adaptations can clarify some inconsistencies in existing models for visual attention, and particularly concerning the interaction between bottom-up and top-down processes. Future research can be concentrated on finding support for the exact location and workings of the IOR mechanism, and in how far different levels of IOR can be present. On the other hand, the current attentional framework needs being converted to a computational vision system and consequently being validated with human experimental data. This conversion is started in cooperation with the department of Electronics and Informatics and tries to implement adapted computer vision models [46].

References

1. Briand, K.A., Klein, R.M.: Is Posner's Beam the same as Treisman's "Glue"?: On the relation between visual orienting and feature integration theory. J. Exp. Psych.: Human Perception and Performance 13, 228–241 (1987)
2. Briand, K.A.: Feature integration and spatial attention: More evidence of a dissociation between endogenous and exogenous orienting. J. Exp. Psych.: Human Perception and Performance 24, 1243–1256 (1998)
3. Itti, L., Koch, C.: Saliency-based search mechanism for overt and covert shifts of visual attention. Visual Research 40, 1489–1506 (2000)
4. Itti, L., Koch, C.: Computational Modelling of Visual Attention. Nature Reviews Neuroscience 2, 194–203 (2001)
5. Wolfe, J.M.: Guided Search 2.0: A revised model of visual search. Psychonomic Bulletin and Review 1, 202–238 (1994)
6. Wolfe, J.M.: Guided Search 4.0: Current progress with a model of visual search. In: Gray, W. (ed.) Integrated models of cognitive systems. Oxford, New York (2006)
7. Henderickx, D., Maetens, K., Soetens, E.: Feature integration and spatial attention: Common processes for endogenous orienting (submitted)
8. Henderickx, D., Maetens, K., Soetens, E.: Inhibition of return: A bottom-up routed attentional process (submitted)
9. Posner, M.I.: Orienting of attention. Quarterly J. Exp. Psych. 32, 3–25 (1980)
10. Remington, R.W.: Attention and saccadic eye movements. J. Exp. Psych.: Human Perception and Performance 6, 726–744 (1980)
11. Godijn, R., Theeuwes, J.: The relationship between exogenous and endogenous saccades and attention. In: Hyönä, J., Radach, R., Deubel, H. (eds.) The Mind's Eyes: Cognitive and Applied Aspects of Eye Movements, pp. 3–26. Elsevier, Amsterdam (2003)
12. Treisman, A.M., Sato, S.: Conjunction search revisited. J. Exp. Psych.: Human Perception and Performance 16, 451–478 (1990)
13. Eriksen, C.W., Hoffman, J.E.: Temporal and spatial characteristics of selective encoding from visual displays. Perception and Psychophysics 12, 201–204 (1972)
14. Yantis, S., Jonides, J.: Abrupt visual onsets and selective attention: Evidence from selective search. J. Exp. Psych.: Human Perception and Performance 10, 601–621 (1984)
15. Posner, M.I., Snyder, C.R.: Facilitation and inhibition. In: Rabbitt, P., Dornick, S. (eds.) Attention and Performance V, pp. 669–682. Academic Press, New York (1975)
16. Jonides, J.: Voluntary vs. automatic control over the mind's eye's movement. In: Long, J.B., Baddeley, A.D. (eds.) Attention and performance IX, pp. 187–203. Erlbaum, Hillsdale (1981)
17. Remington, R.W., Johnston, J.C., Yantis, S.: Involuntary attentional capture by abrupt onsets. Perception & Psychophysics 51, 279–290 (1992)
18. Jonides, J., Yantis, S.: Uniqueness of abrupt stimulus onset in capturing attention. Perception & Psychophysics 43, 346–354 (1988)

19. Klein, R.M.: Perceptual-motor expectancies interact with covert visual orienting under conditions of endogenous but not exogenous control. Canadian J. Exp. Psych. 48, 167–181 (1994)
20. Treisman, A.M., Gelade, G.: A feature integration theory of attention. Cogn. Psych. 12, 97–136 (1980)
21. Posner, M.I., Petersen, S.E.: The attention system of the human brain. Ann. Rev. of Neuroscience 13, 25–42 (1990)
22. Felleman, D.J., Van Essen, D.C.: Distributed hierarchical processing in the primate cerebral cortex. Cerebral Cortex 1, 1–47 (1991)
23. Fox, P.T., Mintun, M.A., Raichle, M.E., Miezen, F.M., Allman, J.M., Van Essen, D.C.: Mapping human visual cortex with positron emission tomography. Nature 323, 806–809 (1986)
24. Folk, C.L., Remington, R.W., Johnston, J.C.: Involuntary covert orienting is contingent on attentional control settings. Journal of Experimental Psychology: Human Perception and Performance 18, 1030–1044 (1992)
25. Folk, C.L., Remington, R.W., Wright, J.H.: The structure of attentional control: Contingent attentional capture by apparent motion, abrupt onset, and color. Journal of Experimental Psychology: Human Perception and Performance 20, 317–329 (1994)
26. Folk, C.L., Remington, R.W.: Selectivity in attentional capture by featural singletons: Evidence for two forms of attentional capture. Journal of Experimental Psychology: Human Perception and Performance 24, 847–858 (1998)
27. Folk, C.L., Remington, R.W.: Can new objects override attentional control settings? Perception & Psychophysics 61, 727–739 (1999)
28. Folk, C.L., Remington, R.W.: Top-down modulation of preattentive processing: Testing the recovery account of contingent capture. Visual Cognition 14, 445–465 (2006)
29. Wurtz, R.H., Goldberg, M.E., Robinson, D.L.: Behavioral modulation of visual responses in the monkey: Stimulus selection for attention and movement. In: Sprague, J.M., Epstein, A.N. (eds.) Progress in psychobiology and physiological psychology, pp. 48–83. Academic Press, New York (1980)
30. Corbetta, M., Miezen, F.M., Schulman, G.L., Petersen, S.E.: A PET study of visuospatial attention. Journal of Neuroscience 13, 1202–1226 (1993)
31. Robinson, D.L., Petersen, S.E.: The pulvinar and visual salience. Trends in Neurosciences 15, 127–132 (1992)
32. LaBerge, D., Buchsbaum, M.S.: Positron emission tomography measurements of pulvinar activity during an attention task. Journal of Neuroscience 10, 613–619 (1990)
33. Posner, M.I., Choate, L., Rafal, R.D., Vaughan, J.: Inhibition of return: Neural mechanisms and function. Cognitive Neuropsychology 2, 211–228 (1985)
34. Petersen, S.E., Fox, P.T., Posner, M.I., Mintun, M., Raichle, M.E.: Positron emission tomographic studies of the cortical anatomy of single-word processing. Nature 331, 585–589 (1988)
35. Gabriel, M.: Functions of anterior and posterior cingulate cortex during avoidance learning in rabbits. In: Uylings, H.B.M., Van Eden, J.P.C., De Bruin, M.A., Feenstra, M.G.P. (eds.) Progress in Brain Research, pp. 467–483. Elsevier, Amsterdam (1990)
36. Amaral, D.G., Price, J.L., Pitkanen, A., Carmichael, S.T.: Anatomical organization of the primate amygdaloid complex. In: Aggleton, J.P. (ed.) The Amygdala: Neurobiological Aspects of Emotion, Memory, and Mental Dysfunction, pp. 1–67. John Wiley & Sons, Inc., New York (1992)

37. Goldman-Rakic, P.S.: Topography of cognition: parallel distributed networks in primate association cortex. An. Rev. of Neuroscience 11, 137–156 (1988)
38. Yantis, S., Jonides, J.: Abrupt visual onsets and selective attention: Voluntary versus automatic allocation. J. Exp. Psych.: Human Perception and Performance 16, 121–134 (1990)
39. Clark, J.J.: Spatial attention and latencies of saccadic eye movements. Vision Research 39, 585–602 (1999)
40. Posner, M.I., Rafal, R.D., Choate, L., Vaughan, J.: Inhibition of return: Neural basis and function. Cognitive Neuropsychology 2, 211–228 (1985)
41. Rafal, R.D., Calabresi, P.A., Brennan, C.W., Sciolto, T.K.: Saccade preparation inhibits reorienting to recently attended locations. Journal of Experimental Psychology: Human Perception and Performance 15, 673–685 (1989)
42. Ungerleider, L.G., Mishkin, M.: Two cortical visual systems. In: Ingle, D.J., Goodale, M.A., Mansfield, R.J.W. (eds.) Analysis of Visual Behavior, pp. 549–586. MIT Press, Cambridge (1982)
43. Milner, A.D., Goodale, M.A.: The visual brain in action. Oxford University Press, Oxford (1995)
44. Maunsell, J.H.R., Newsome, W.T.: Visual processing in monkey extrastriate cortex. Annual Review of Neuroscience 10, 363–401 (1987)
45. DeSchepper, B., Treisman, A.: Visual memory for novel shapes: implicit coding without attention. J. Exp. Psych.: Learning, Memory, and Cognition 22, 27–47 (1996)
46. Geerinck, T., Sahli, H., Vanhamel, I., Enescu, V.: Modeling attention and perceptual grouping to salient objects. In: Paletta, L., Tsotsos, J.K. (eds.) Proceedings of the Fifth International Workshop on Attention in Cognitive Systems, Santorini, Greece, pp. 235–248 (May 2008)

Relative Influence of Bottom-Up and Top-Down Attention

Matei Mancas

Engineering Faculty of Mons (FPMs)
31, Bd. Dolez, 7000 Mons, Belgium
Matei.Mancas@fpms.ac.be

Abstract. Attention and memory are very closely related and their aim is to simplify the acquired data into an intelligent structured data set. Two main points are discussed in this paper. The first one is the presentation of a novel visual attention model for still images which includes both a bottom-up and a top-down approach. The bottom-up model is based on structures rarity within the image during the forgetting process. The top-down information uses mouse-tracking experiments to build models of a global behavior for a given kind of image. The proposed models assessment is achieved on a 91-image database. The second interesting point is that the relative importance of bottom-up and top-down attention depends on the specificity of each image. In unknown images the bottom-up influence remains very important while in specific kinds of images (like web sites) top-down attention brings the major information.

Keywords: Visual attention, saliency, bottom-up, top-down, mouse-tracking.

1 Introduction

The aim of computational attention is to automatically predict human attention on different kinds of data such as sounds, images, video sequences, smell or taste, etc... This domain is of a crucial importance in artificial intelligence and its applications are numberless from signal coding to object recognition. Intelligence is not due only to attention, but there is no intelligence without attention.

Attention is also closely related to memory through a continuous competition between a bottom-up approach which uses the features of the acquired signal and a top-down approach which uses observer's a priori knowledge about the observed signal. While eye fixations add novel information to both short-term and long-term memory [1], long-term spatial context memory is able to modify visual search [2]. In this paper, a new model of bottom-up attention and a way to build top-down models of attention are proposed. An assessment of this approach is achieved which leads to a discussion on the relative importance of bottom-up and top-down influence.

In the next section, a state of the art in computational attention is achieved. The third section presents an original bottom-up computational attention which highlights the regions within an image which remain rare during the forgetting process. The fourth section proposes a way of building top-down models which contain the mean behavior of the observers for specific images. Section 5 achieves a computational attention algorithm assessment on a 91-image database. Finally, this section is followed by a discussion and a conclusion.

L. Paletta and J.K. Tsotsos (Eds.): WAPCV 2008, LNAI 5395, pp. 212–226, 2009.
© Springer-Verlag Berlin Heidelberg 2009

2 Computational Attention: A State of the Art

The result of attention algorithms is called an "attention map" which is a mono-dimensional intensity matrix with the same size as the input image and which provides higher intensities for the most important areas of the initial image (visual field). If only bottom-up attention is taken into account this attention map is often called the "saliency map" of the input signal.

The number of computational models has recently exploded as a confirmation of the maturity of the knowledge acquired within the biological, psychological and neuroscience domains. Several classifications of these methods are obviously possible, and most of them have similar philosophies, however it is possible to distinguish two main ideas. Attention may be due to:

- Local properties (a feature saliency depends on its neighborhood)
- Global properties (a feature saliency depends on the whole visual field).

If biological evidences supporting the local approaches are numerous, global approaches are for instance less well biologically motivated. This situation is normal as the local behavior of the cells on their classical receptive field (CRF) is obvious. Nevertheless, recent experiments in visual attention [3], [4] brought interesting confirmations for a global integration of features information all over the visual field. This is possible thanks to the impressive neuronal network which includes an important amount of "horizontal cells" which connect more or less directly the cells from the whole visual field.

2.1 Mostly Local Methods

In 1998, Itti et al. ([5], [6], [7]), set up the most well-known computational attention model. Based on the Koch and Ullman model [8], Itti proposed the extraction of three main features: luminance, chrominance, and orientation. These features are processed in parallel and then fused within a single saliency map.

Milanese et al. ([9], [10]) proposed an attention approach also based on the seminal architecture of Koch and Ullman. They added two more features which are contours amplitude and curvature. The normalization step is done by using Gaussian filtering and gradient descent-based relaxation before getting the mean of the maps.

Chauvin et al. [11] used the Koch and Ullman architecture with Gabor filtering to get multi-resolution information. Additional computations reinforcing collinear and longer contours are also added. This model only deals with the luminance features, avoiding the difficult normalization step, but loosing important color information.

Petkov et al. [12] proposed a lateral inhibition technique to distinguish object contours from image texture. Le Meur et al. ([13], [14]) achieved a computational model of visual attention which is one of the closest to the local processing biological reality within the human visual system. Also based on the Koch and Ullman architecture, it integrates biological data for intermediate maps data fusion.

2.2 Mostly Global Methods

Mudge et al. [15] suggested as early as 1987 that object components saliency may be inversely proportional to their occurrence within the image. Osberger and Maeder [16] used a segmentation approach to separate the image into several homogenous areas. Five features were used in assigning a relative importance to the segmented areas. The problem of this kind of approaches is that errors within the segmentation may induce errors in the attention map.

Walker et al. [17] suggested that saliency may be related to the probability that a feature has to be misclassified with all the other features within an image.

Oliva et al. ([18], [19]) had a similar approach to Mudge et al. by stating that attention should be inversely proportional to the existence probability of a pixel. They modeled this probability with a Gaussian and used multi-resolution wavelet decomposition. Results seem similar to Itti's model as compared to eye tracking results. An interesting fact is that results are better than Itti's model if additional top-down information is used.

Bruce and Jernigan [20] integrated this idea by turning it into an information theory approach within the Koch and Ullman architecture. They afterwards [21] used ICA (Independent Component Analysis) to compare local features (local random patches of the image) in an image patches database obtained from the current image but also from other images.

Liu et al. [22] used image segmentation as Osberger and Maeder, but the mean shift [23] technique let it provide a more robust segmentation. They also assumed that centered regions may have higher attention scores. In section 4 of this paper it will be shown that this assumption is verified only in the case of natural scene images!

Itti and Baldi [24] also published a probabilistic approach of surprise based on the Kullback-Leibler divergence which is the energy of the so-called "net surprisal" within the information theory. The idea is that attention is due to a more or less important difference between what was expected to happen and the actual observation. This method has been integrated into Itti's model architecture and it provides better results compared to the original approach.

Stentiford [25] proposed a method related with Walker's ideas, but he defined no specific feature. Random pixel neighborhoods (forks) are directly compared and they are declared as matching if the distance between the two neighborhoods is below a threshold. If few matches are observed, the pixel is assigned with a high saliency score. The method provides very interesting results and its main advantage is to remain very general. It takes into account intensity, colors, directions and shapes mostly to smaller scales.

Boiman and Irani ([26], [27]) used comparisons between gradient-based patches of different sizes to define occurrence probabilities. One of the main originalities of this method is in the fact that not only different patches from the images are compared with other patches in the same image or in a database, but also the relative patches' positions were taken into account.

The author proposed ([28], [29], [30]) a global rarity approach of attention but not much local information was taken into account.

The approach proposed here is based on the fact that visual attention is the first filter which selects regions in an image which may be interesting to memorize. This fact implies that during the forgetting process, rare regions are kept in mind while the others are forgotten. The proposed model will use both global rarity and local contrast information and thus it has properties from both local and global approaches.

3 Bottom-Up Attention: An Unsupervised Signal-Based Approach

This section describes a bottom-up attention approach which could also be seen as an unsupervised attention. Bottom-up attention uses the acquired image characteristics to predict its important regions and acts like a gate to memory. This model is somehow based on Edgar Alan Poe's proposition: "observing attentively is remembering clearly". Unsupervised attention is thus very important in remembering and it is able to keep in mind the details or rare regions within the image. Without attention, these important details are forgotten which implies a loss of crucial data.

When performing a remembering task about an already visited place for example, people remember a rough image about this place. The process of forgetting may be modeled by a low-pass filtering whose kernel size increases in time. Here, a set of six low-pass filters with increasing kernel sizes is used for each grey level of the image. The number of grey levels is reduced to 11 to speed up computation and avoid noise. The size of the largest low-pass filter kernel is chosen to be close to the half of the image. If the original image is larger or smaller than the largest filtering kernel size, it is resized to better fit the scale decomposition.

This idea is illustrated in Fig. 1 where the sky (big upper rectangle) and a pool (small rectangle in the middle) have the same grey level (let's say "blue"). At the higher resolution (top row on Fig. 1), two pixels (one in the middle of the sky and the other one in the middle of the pool) have the same global occurrence which is equal to the number of "blue" pixels. When going from top images to bottom images in Fig. 1, low-pass filter kernels sizes (neighborhood sizes) are larger, thus the images are forgotten more and more. The occurrences of the two pixels have different behaviors (plots of the left column: sky; plots on the right column: pool). If the pixel within the sky has a slowly decreasing occurrence, the pool pixel's occurrence decreases very fast when larger and larger neighborhoods are taken into account (larger low-pass kernels). The pool pixel has an occurrence which gets rapidly very small while the sky pixel keeps a higher occurrence even when taking into account larger neighborhood sizes.

In order to quantify the behavior for each pixel, the sum on the scale space is used. This sum can be visualized in Fig. 1 on the right and left columns as the area behind the occurrence variation plots function of the neighborhood size. The occurrence probability of a pixel is obtained by the normalization of this sum and the self-information represents the attention score for the pixel (Eq. 1).

$$Attention(I_j) = -\log\left(\frac{1}{S \times Card(I_j)} \sum_{k=1}^{S} n_k\right) \tag{1}$$

In Equation 1, n_k is the occurrence value of the current pixel within the kth resolution level. In the current implementation there are seven different resolutions and the one corresponding to $k=1$ means the grey level is unfiltered. S is a constant which is equal to the total number of resolutions (here $S=7$). I_j is the jth grey level of the image I and $Card(I_j)$ is its cardinality.

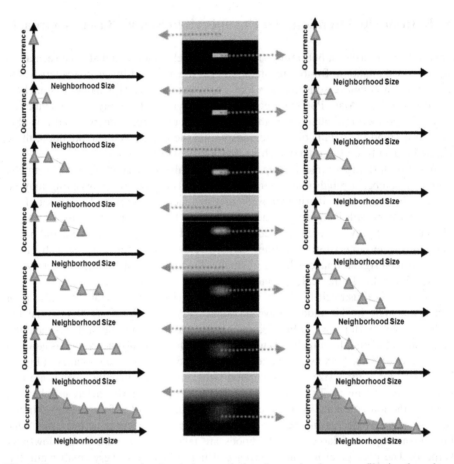

Fig. 1. From top to down: the decreasing resolution (increasing low-pass filtering kernel or neighborhood size) of an initial image grey level, From left to right: the occurrence behavior of a pixel in the upper red rectangle ("the sky") during the forgetting process; an image grey level; the occurrence behavior of a pixel in the lowest red rectangle ("the pool") during the forgetting process

In the current implementation, instead of simply using grey levels, their contrast maps are used. These maps are obtained as follows: for the grey level j (j is set between 1 and 11 in the current implementation), the pixels equal to j are assigned with the value 1 while pixels different from j are assigned with a value between 0 and 1. This value will be close to 1 if the pixel has a similar value with j and close to 0 if this pixel has a grey level, very different from j.

The color images are handled within an opposition color system. For each of the components (luminance, red-green opposition, blue-yellow opposition) a separate attention map is computed: the final map is obtained by adding the maps with a higher weight on the luminance which contains more information.

4 Top-Down Attention: A Supervised Application-Driven Approach

While a bottom-up approach uses signal characteristics to achieve attention computation, the top-down approach mainly uses feedbacks from the memory (a priori knowledge) and it depends on the task or the application to be achieved. Top-down attention can be seen as a supervised attention. In this section, a top-down approach for still images is proposed. The idea is to model the observers' behavior depending on the kind of images they look at.

Observers' behavior can be modeled by using eye-tracking or other alternative methods such as mouse-tracking to detect their gaze path. The mean of the gaze path of several observers is called a priority map and it highlights, for one image, the areas where the mean of a set of observers mostly looks as it is shown in Fig. 2.

Fig. 2. Left: original image, right: a priority map obtained by mouse-tracking

A top-down model can be achieved by using the mean of the priority maps obtained for a specific set of images (images with common meaning). Three sets of images [31] (set1, set2 and set3) were used within these tests to build three different top-down models:

- The first image set is made from 26 natural scene images. Some examples are available in Fig. 4, first row.
- The second image set is made from 30 various advertisement images. Well-known trade marks were chosen as they have a huge advertising presence. Some examples are available in Fig. 5, first row.
- The third image set is made from 35 various web sites. The web sites of the 12 candidates to the French presidential election of 2007 are analyzed along with university and lab web sites, institutional and government web sites. Some commercial web sites have also been added. These media do not intend to provide the same informational content that is why it is interesting to see if there is a common attentional behavior to all these websites. Some examples are available in Fig. 6, first row.

It is important to highlight the fact that a top-down model built in that way needs two main requirements to be meaningful:

- The first one is about the number of observers who provide their mouse track paths which should be high enough to get a realistic observer mean. Here, 40 to 60 observers' mouse paths per image were recorded. There were neither advertisement or web experts nor a specific age or gender class of observers: they can reasonably be considered as general public.
- The second requirement is about the homogeneity of the image set. The more the set of images is specific, the more the top-down model is accurate.

Fig. 3 displays the three top-down models from left to right: the sets presenting natural images, advertisements and web sites. For the natural scene images, the mean priority map is mostly centered and it oddly looks like a centered Gaussian. The two other models are quite similar: high scores are detected in the top-left corner of the image decreasing towards the center. Nevertheless, the models used for advertisements and web sites also have some differences. Fig. 3 shows that the advertisement model is less selective than the web sites one: structures on its center are also quite well highlighted.

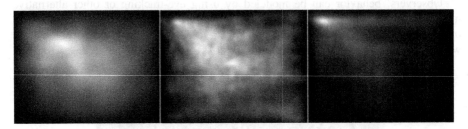

Fig. 3. Left to right: top-down models for a set of natural images, advertisements, web sites

The web sites model is a typical structured document model. The natural images model is typical of unknown unstructured images. The advertisements model seems to be a mix between these two extreme models. Its structure is close to a structured document one (human contribution is high), but it also covers the central areas of the image and the opposite corners where logos may often be found. This experiment shows how observers' attention behavior is different depending on the set of images.

To mix the bottom-up maps with the top-down maps a simple multiplication between those two normalized maps is used here.

5   Computational Attention Evaluation

5.1   Algorithms Comparison: A Qualitative Approach

The bottom-up attention model proposed in section 3 and the Itti's reference bottom-up saliency map [5] were compared from a qualitative point of view on the 91-images database. The top-down models proposed in the previous section (Fig. 3) were also used for this comparison. As discussed in section 5.5, comparisons between these two methods are not very easy as the nature of the attention map is not the same. Nevertheless, for most of the images, the proposed method seems to perform better for the bottom-up algorithm alone but also when it is used along with the corresponding top-down model: Itti's model often overestimates some spatial orientations and local contrast cues.

Figs. 4, 5 and 6 show for each set of the database three examples of images. The original images are presented in the first row. It should be known that those images were initially color images and they were resized in order to better fit into the figures. They were only chosen for the pertinence of the test results and not for their content.

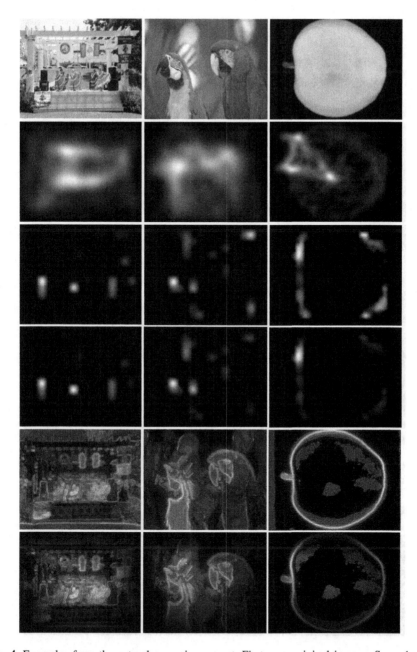

Fig. 4. Examples from the natural scene images set. First row: original images, Second row: mouse-tracking priority maps (gold standard here), Third row: Itti's method (bottom-up), Fourth row: Itti's method (bottom-up + corresponding top-down model), Fifth row: proposed method (bottom-up), Sixth row: proposed method (bottom-up + corresponding top-down model).

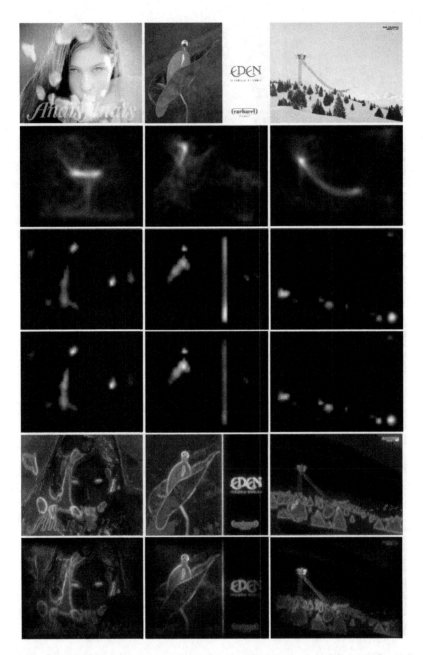

Fig. 5. Examples from the advertisement images set. First row: original images, Second row: mouse-tracking priority maps (gold standard here), Third row: Itti's method (bottom-up), Fourth row: Itti's method (bottom-up + corresponding top-down model), Fifth row: proposed method (bottom-up), Sixth row: proposed method (bottom-up + corresponding top-down model).

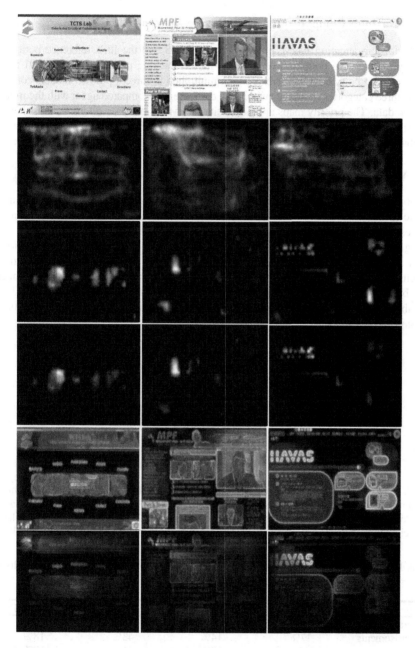

Fig. 6. Examples from the web sites images set. First row: original images, Second row: mouse-tracking priority maps (gold standard here). Third row: Itti's method (bottom-up), Fourth row: Itti's method (bottom-up + corresponding top-down model), Fifth row: proposed method (bottom-up), Sixth row: proposed method (bottom-up + corresponding top-down model).

5.2 Algorithms Comparison: A Quantitative Approach

The bottom-up attention model proposed in section 3 and the Itti's reference bottom-up saliency map [5] are both compared to the mouse-tracking priority maps of a 91-image database. Both bottom-up methods are then multiplied to three top-down models concerning the three categories of images within the database and compared again with the mouse-tracking data. The priority maps are the results of several low-pass filterings, thus they are very smooth. In order for all compared algorithms and priority maps to have the same spatial characteristics (smooth areas) the algorithm proposed in section 3 is low-pass filtered.

5.3 Algorithms Comparison: Bottom-Up Information Only

To obtain quantitative results, the classical absolute value of the linear correlation metric is used here. The correlation value goes from 0 (no similarity between the images) to 1 (there is a linear relationship between them).

Fig. 7 displays the correlation coefficient between the mouse-tracking priority maps and both Itti's bottom-up algorithm (dotted plot) and the proposed bottom-up algorithm (solid plot) for all images in the database. Very often, the tested images have better correlation coefficients for the proposed algorithm than for Itti's one.

Table 1 shows the mean and standard deviation of the correlation coefficients for the three sets of images.

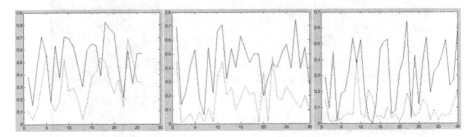

Fig. 7. Correlation coefficient between the priority maps and both the bottom-up algorithm of Itti (dotted plot) and the proposed bottom-up algorithm (solid plot). From left to right: natural scene database, advertising database, web sites database. The Y axis represents the correlation coefficient with the mouse-tracking results and the X axis the image number from the database.

Table 1. Bottom-up linear correlation mean (MEAN) and standard deviation (STD) results for the three sets of images

Image set	Itti MEAN	Itti STD	Mancas MEAN	Mancas STD
Natural Images	30%	18%	52%	19%
Advertisements	17%	14%	43%	18%
Web Sites	9%	10%	31%	19%

5.4 Algorithms Comparison: Both Bottom-Up and Top-Down Information

In this section, the same top down models (those proposed in section 4 and displayed in Fig. 3) were added to Itti's and to the proposed (Mancas) bottom-up algorithms.

High results improvements may be observed (Fig. 8) compared to the previous section where no top-down influence was taken into account. Moreover, even with the same top-down models, one may see that the bottom-up model remains very important as results of the proposed bottom-up method are better in terms of linear correlation than those of Itti's bottom-up model for 90 images on 91.

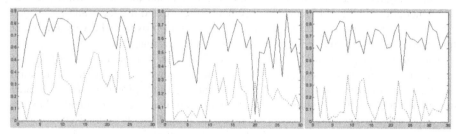

Fig. 8. Correlation coefficient between the priority maps and both the bottom-up and top-down attention algorithms based on Itti (dotted plot) and on Mancas (solid plot). From left to right: natural scene database, advertising database, web sites database. The Y axis represents the correlation coefficient with the mouse-tracking results and the X axis the image number from the database.

Table 2 summarizes the mean and standard deviation of the correlation coefficients for the three sets of images. A simple comparison with Table 1, where no top-down information was used shows the importance of the top-down step in attention.

Table 2. Bottom-up & top-down linear correlation mean (MEAN) and standard deviation (STD) results for the three sets of images

Image set	Itti MEAN	Itti STD	Mancas MEAN	Mancas STD
Natural Images	34%	17%	74%	12%
Advertisements	18%	14%	53%	16%
Web Sites	13%	11%	69%	9%

5.5 Algorithms Comparison: A Discussion

The bottom-up influence is higher for natural scene images than for websites images for example. For both Mancas and Itti methods, the bottom-up attention alone (Table 1) provides the best results for natural scene images, while this score decreases with the advertisement set and even more with the web sites set.

Moreover, the results of Table 2 show that the 74% of correlation for natural images (Mancas method) is due to 52% bottom-up and 22% top-down. On the other side, the result of 69% of correlation for the web site images is due to only 31% of bottom-up and to 38% top-down influence. A similar behavior can also be detected on the figures of the Itti method of Table 2.

A very interesting conclusion of these observations is that the more one knows about an image, the higher the top-down influence part will be. On the other side, for an unknown image, the bottom-up attention mechanism will be very important. Thus, if the role of top-down information in attention is always very important, its part in the attention process depends on the amount of knowledge that a mean observer may have on a given kind of images.

Nevertheless, advertisements score enhancement between Table1 and Table 2 is lower than expected: as advertisements are a mix between unstructured (natural scenes) and structured (web sites) documents, the influence of the top-down attention should be higher than for natural scene images. This is not the case because the top-down model used here only includes document structure and not faces and text which are also very powerful top-down stimuli. These stimuli proved to be very important for advertisements where faces and text are often very present.

The correlation results of Table 1 and Table 2 also need some remarks. The use of the linear correlation may not be the best metric to compare computational attention algorithms and other distances could be taken into account. Moreover, Itti's saliency map which focuses very highly on precise areas in an image may be penalized by the use of the linear correlation coefficient. It is thus quite difficult to precisely compare these two methods which have different behavior as there is no standard method for attention algorithms assessment. However, a precise analysis of both qualitative and quantitative results on the overall database shows that the proposed bottom-up algorithm outperforms the bottom-up algorithm proposed by Itti.

The purpose of this section was to show that the correlation coefficients between the mouse-tracking priority maps and the proposed algorithm become very interesting and they can be considered as a quite good approximation of human attention. These correlation figures demonstrate that the use of attention to predict human gaze makes sense if both bottom-up and top-down information are used.

6   Conclusion

A bottom-up or unsupervised computational attention algorithm is presented which performs better than Itti's reference model on the test database. However, several improvements should be added to this algorithm, mainly to handle spatial orientations.

A top-down or supervised attention model based on the mean of the eye-tracking or mouse-tracking priority maps was also proposed. It proved to highly increase the results of the bottom-up algorithms and to finally provide a good approximation of human gaze. Other top-down influences should also be added to improve the results as face or text detection. Faces and text are known as very informative and that is why they represent very important top-down influences especially if there are few faces or few text within the images.

The encouraging results presented here confirm the more and more widely accepted idea that the automatic prediction of human attention for still images becomes quite accurate if both bottom-up and top-down information is used while bottom-up information alone remains insufficient. An issue in computational attention is the lack of a standard assessment method and database to really prove the pertinence of those approaches in predicting human attention.

An interesting point was found about the relative importance of bottom-up and top-down influences: the bottom-up mechanism is very important for new images, while for structured document where people are used with, the top-down influence is higher than the bottom-up one. These results, which can be seen in sections 5.3 and

5.4, show that there is a complex relationship between bottom-up and top-down attention. On one side, bottom-up attention aims in learning which areas of an image are the most relevant: this process is mainly used for new images and situations. On the other side, once some situations are learnt using bottom-up attention, top-down attention uses this information to select some areas of the current image by inhibiting those where there are very few chances to find relevant information.

Bottom-up and top-down attention interaction aims in optimizing reactions to both novel and already experienced situations.

Acknowledgements

This work was achieved in the framework of the Translogistic project which is funded by the Walloon Region, Belgium.

References

1. Hollingworth, A.: Constructing visual representations from natural scenes: The roles of short- and long-term visual memory. Journal of Experimental Psychology: Human Perception and Performance 30, 519–537 (2004)
2. Chun, M.M., Jiang, Y.: Implicit, long-term spatial context memory. Journal of Experimental Psychology: Learning, Memory, & Cognition 29, 224–234 (2003)
3. Desimone, R.: Visual attention mediated by biased competition in extrastriate visual cortex. Phil. Trans. R. Soc. Lond. B 353, 1245–1255 (1998)
4. Boynton, G.M.: Attention and visual perception. Current Opinion in Neurobiology 15, 465–469 (2005)
5. Itti, L., Koch, C., Niebur, E.: Model of saliency-based visual attention for rapid scene analysis. IEEE Trans. on Pattern Analysis and Machine Intelligence 20(11), 1254–1259 (1998)
6. Itti, L., Koch, C.: A saliency-based search mechanism for overt and covert shifts of visual attention. Vision Research 40, 1489–1506 (2000)
7. Itti, L., Koch, C.: Computational modeling of visual attention. Nature RevNeuroscience 2(3), 194–203 (2001)
8. Koch, C., Ullman, S.: Shifts in selective visual attention: towards the underlying neural circuitry. Human Neurobiology 4(4), 219–270 (1985)
9. Milanese, R., Bost, J.M., Pun, T.: A bottom-up attention system for active vision. In: ECAI 1992, 10th European Conference on Artificial Intelligence, pp. 808–810 (1992)
10. Milanese, R.: Detecting salient regions in an image: from biological evidence to computer implementation. PhD Thesis, University of Geneva (1993)
11. Chauvin, A., Herault, J., Marendaz, C., Peyrin, C.: Natural scene perception: visual attractors and image processing. In: 7th Neural Computation and Psychology Workshop (2000)
12. Petkov, N., Westenberg, M.A.: Suppression of contour perception by band-limited noise and its relation to non-classical receptive field inhibition. Biological Cybernetics 88, 236–246 (2003)
13. Le Meur, O.: Attention sélective en visualisation d'images fixes et animées affichées sur écran: Modèles et évaluation des performances – Applications. PhD Thesis, University of Nantes (2005)

14. Le Meur, O., Le Callet, P., Barba, D.: A spatio-temporal model of bottom-up visual selective attention: description and assessment. Vision Research (2007)
15. Mudge, T.N., Turney, J.L., Volz, R.A.: Automatic generation of salient features for the recognition of partially occluded parts. Robotica 5, 117–127 (1987)
16. Osberger, W., Maeder, A.J.: Automatic identification of perceptually important regions in an image. In: 14th IEEE Int. Conference on Pattern Recognition (1998)
17. Walker, K.N., Cootes, T.F., Taylor, C.J.: Locating Salient Object Features. In: British Machine Vision Conference (1998)
18. Oliva, A., Torralba, A.: Modeling the shape of the scene: a holistic representation of the spatial envelope. International Journal of Computer Vision 43(3), 145–175 (2001)
19. Oliva, A., Torralba, A., Castelhano, M.S., Henderson, J.M.: Top-down control of visual attention in object detection. In: IEEE International Conference on Image Processing (2003)
20. Bruce, N., Jernigan, E.: Evolutionary design of context-free attentional operators. In: Proc. of the IEEE International Conference on Image Processing (2003)
21. Bruce, N., Tsotsos, J.K.: Saliency Based on Information Maximization. In: Proc. of the Neural Information Processing Systems (2005)
22. Liu, F., Gleicher, M.: Video Retargeting: Automating Pan-and-Scan. ACM Multimedia (2006)
23. Comaniciu, D., Meer, P.: Mean Shift: A Robust Approach toward Feature Space Analysis. IEEE Trans. Pattern Analysis Machine Intell. 24(5), 603–619 (2002)
24. Itti, L., Baldi, P.: Bayesian Surprise Attracts Human Attention. In: Advances in Neural Information Processing Systems (NIPS 2005), vol. 19, pp. 1–8. MIT Press, Cambridge (2006)
25. Stentiford, F.W.M.: An estimator for visual attention through competitive novelty with application to image compression. In: Picture Coding Symposium, pp. 25–27 (2001)
26. Boiman, O., Irani, M.: Detecting Irregularities in Images and in Video. In: International Conference on Computer Vision (ICCV) (2005)
27. Boiman, O., Irani, M.: Similarity by Composition. In: Neural Information Processing Systems (NIPS) (2006)
28. Mancas, M., Mancas-Thillou, C., Gosselin, B., Macq, B.: A rarity-based visual attention map - application to texture description. In: Proc. of IEEE International conference on Image Processing (ICIP) (2006)
29. Mancas, M., Gosselin, B., Macq, B.: A Three-Level Computational Attention Model. In: Proceedings of ICVS Workshop on Computational Attention & Applications (WCAA) (2007)
30. Mancas, M., Gosselin, B., Macq, B.: Perceptual Image Representation, EURASIP Journal of Image and Video Processing, Article ID 98181, doi:10.1155/2007/98181 (2007)
31. Validattention website,
    http://tcts.fpms.ac.be/~mousetrack/pageAccueil.php?langue=en

Towards Standardization of Evaluation Metrics and Methods for Visual Attention Models

Muhammad Zaheer Aziz and Bärbel Mertsching

GET LAB, Universität Paderborn, 33098 Paderborn, Germany
{last name}@upb.de
http://getwww.uni-paderborn.de

Abstract. Every field of science requires standardization of metrics and measurement methods for detecting true advancement in research. Efforts on computational models of visual attention models have increased in the recent years and now it is important to have standard measuring techniques in this area in order to avoid undue deceleration in its progress. This paper performs a review of the evaluation techniques used by different researchers in the field and brings them in an organized structure. Further methods and metrics are also proposed that would lead to more objective and quantitative evaluation of the attention models.

1 Introduction

Every field of science has a way for objectively evaluating the outcome of processes by representing the quantities with magnitudes measured in suitable units. The magnitudes are obtained through well-defined measurement methodologies and, when needed, some standard calculations. Despite having an age of a couple of decades, the field of attention modeling still lacks standard metrics and methods for measuring and evaluating outcome of a model. The number of emerging models has increased in the recent years and each model claims to be performing better than the contemporary ones in one aspect or the other. Such claims may not always reflect true advancement because the evaluation method could be measuring a characteristic that either has trivial relevancy to the actual need of progress in the field or leads to a biased comparison. It is now an appropriate time to devise standards for benchmarking performance of the models in order to prevent the research in this area from iteratively circling or adopting an extremely low pace of true progress.

This paper is an attempt towards designing standard metrics and methods for evaluation of visual attention models using which the degree of success for output of a model could be measured and the model's results could be compared to some benchmark data. Due to complex and multi facet nature of attention's output there exist many aspects that can be evaluated. Standardization is needed for each aspect for a fair and justifiable evaluation. In this discussion, the evaluation techniques used in available work on visual attention are reviewed to collect them together, the factors that can be evaluated in a computational model of visual

L. Paletta and J.K. Tsotsos (Eds.): WAPCV 2008, LNAI 5395, pp. 227–241, 2009.

attention and the relevant yardsticks are analyzed, and further proposals are made to make the evaluation mechanism quantifiable. We prefer the measures that obtain the quantity of the efficiency or performance in the range between 0 and 1 due to ease of comparison using such values and their scalability.

The research on the phenomenon of attention has a history of two and a half centuries. Results of experiments reported in literature on psychophysics and neurobiology provide valuable ground-truth data in different formats, such as activity spots and scan paths, to build computational models of attention and to verify their output. Performance evaluation of an attention model requires answers to many questions. The first question is to decide the aspect of attention (AA) that should be evaluated. For example, evaluation of an attentive visual search has different criteria as those for free viewing. The second question requires to decide the evaluation aspect (EA), for example the aspect of evaluation could be validation of the model's output by comparing it with results of natural attention or to test the model's competence by measuring the level of success in achieving a predefined task. The output of attention has various styles, in which saliency maps, fixated locations, and scan path sequences are commonly known. The third question asks that which output format (OF) should be used to evaluate a particular aspect of attention, for example in order to measure a model's efficiency in discovering salient locations it may be suitable to use fixation spots obtained in a given time rather than a scan path recorded for an unrestricted period. Each of these output formats conveys information that can be interpreted in different ways, for example, a fixation map can be viewed in perspective of FOA locations, order of fixations, or both. In this regard, the fourth question is to decide the perspective of results (PR) that suits best to the evaluation aspect. Every perspective of observing results of an attention model will need a special metric and measurement method, hence, the fifth question requires to associate a proper measurement method (MM) to the chosen perspective of results and also to define methods to extract readings from the model's output required for that measurement.

Yet there is another question that demands selection of proper benchmark data with which the model's output should be compared. We will not address this question in this discussion except for the recommendation that nature of the selected benchmark data should match the task given to the artificial attention mechanism under evaluation. Answers for the other questions will be investigated here assuming that the reference data suites well to the model's active behavior.

2   A Survey of Available Techniques

In this section we perform a survey of existing techniques in which useful tools and metrics for evaluation of attention models under different contexts will be collected together. These techniques will be analyzed in context of their methodology and measurement metrics to obtain a comprehensive synopsis of available evaluation tools. An analysis is performed in the next section.

In the evaluation scheme used in [1], a comparison of locations is performed with results of spatial frequency output maps in a subjective manner. The metrics used for measuring acceptability of a model's output are number of fixations before attending a target (salient objects marked by human subjects), number of fixations incident on objects of human interest (such as faces, flags, persons, buildings, or vehicles), and variation in these quantities after introducing increasing amount of artificial noise in the input. Acceptability of fixations is decided by human subjects.

For the evaluation scheme used in [2] subjects were required to free-view naturalistic and artificial images while their eye movements were recorded and the resulting fixation locations were compared with the saliency maps produced by the model. Each salience map was linearly normalized to have zero mean and unit standard deviation. Next, the normalized saliency values were extracted from each point corresponding to the fixation locations along a subject's scan path and the mean of these values, named as normalized scan path salience (NSS), was taken as a measure of the correspondence between the salience map and the scan path. NSS values greater than zero suggest a greater correspondence, a zero indicates no correspondence, and negative values indicate an anti-correspondence between fixations and model-predicted salient points.

According to [3] an attention model should be able to explore all locations that can be of interest for a human observer. They get the salient objects marked by human subjects and then measure that in how many fixations these objects get covered by the model. They compare the model's performance on the basis of false fixations before focusing on the first required object and false fixation before covering 50%, 75%, and 100% of the target locations.

The main objective of the work in [4] is to adapt large images into videos for small size displays like mobile phones such that all the important locations for human vision are covered in the small video representing the image. Evaluation is done on locations as well as sequence of fixations by including queries in the questionnaire for human evaluators. The questions ask whether all the important Regions of Interest (ROIs) in the image are focused in the video clip and if the display order of the focused ROIs in the video clip with the order the user would focus on the same ROIs in the given image. A similar but simpler scheme for evaluating down sampled images is proposed in [5].

The scheme given in [6] makes a comparison of ROI clusters. The compared two sets of ROIs are clustered using a distance measure derived from a k-means pre-evaluation. Any two ROIs in the compared sets that are closer than a certain distance are considered as coincident and those far apart than this distance as non-coincident. The value of similarity metric S_p, representing common ROIs in the two sets, is obtained through the following process. In order to compare the sequence of fixations each ROI from ground truth is labeled with a separate letter and these letters are concatenated in the order of appearance of the corresponding ROI to form a "ground-truth-string". A similar string is created for the output of the attention model. All the coincident ROIs are labeled with the same alphabetic character. Calculating the cost of transforming one string into the other compares the two obtained strings. Costs are defined for insertion, deletion and substitution of letters. The minimum costs of this transformation are computed using dynamic programming approach.

The method discussed in [7] extends the above mentioned technique [6] by arguing that the said method has a limitation of defining two regions of interest with equal importance as one ROI has to be always preferred over another in order to set up the ground truth and its labeling order. The proposed hybrid approach claims to be able to handle situations of order uncertainties in ROIs. They assign numbers to the ROIs according to the relative order and store the strings in a matrix. These operations are repeated and average of the iterations is obtained in a resultant matrix. Such matrices are created for ground-truth and the test case. Now the magnitude of the normalized cross-correlation of the two matrices gives the measure of similarity between the sets of ROIs.

The evaluation scheme presented in [8] proposes that a system should be able to attend to the same features in a scene, whether or not the scene has been translated, rotated, reflected or scaled. They quantify the performance of an attention system through two measures. The first looks for gross error rate, that records the percentage of fixations in the test image that are not within a threshold radius of any fixation in the transformed image, once the geometric transformation is compensated for. The second measure is a form of the Hausdorf distance metric that measures positional noise. Having the Hausdorf distance $h(A, B) = median(min\|a - b\|^2)$, the positional noise is measured as $max(h(A, B), h(B, A))$ where A is the set of fixations from the original test image and B is the set of compensated fixations from the transformed test image. As long as fewer than half the locations are outliers, this statistic reflects the positional noise between the two sets of fixations.

The work in [9] proposes to compare the sequence of ROIs identified by an attentional algorithm to those foveated by human observers. Two methods of temporal analysis are presented, namely, head-based and time-based. Head-based analysis is used to locate sequences of still images, from the live input, where the head is stable (but the eyes may not be). Based on estimates of head stability, this analysis technique presents a still image to the attentional model for variable periods of viewing time. The scanpath generated from the analysis of head movements is compared with the scanpath followed by the given model using the string editing technique proposed in [6]. The time-based analysis approach assumes a constant frame rate of 10 fps hence the attentional model is constrained to locate ROIs within a constant time period of 100 ms, hence human fixations and ROIs from the model are compared over frames collected every 100 ms. To allow comparison between ROIs from the model and human fixations in both approaches, human fixations are identified via velocity based analysis of eye movements over the same input given to the attentional model.

The model of [10] have used a method to compare saliency maps based upon linear correlation coefficient, which measures the strength of a linear relationship between two variables. This measure allows to compare two variables by providing a single scalar value between 1 and −1 where the correlation close to ±1 represents an almost perfectly linear relationship between the two variables. For p and h being the fixation density maps by the model and humans respectively, the correlation constant $cc(p, h)$ is computed as $cc(p, h) = cov(p, h)/(\sigma_p \sigma_h)$, where $cov(p, h)$ is the covariance value between p and h while σ_p and σ_h are the standard deviation for p and h respectively.

The method proposed in [11] names its metric as score-s. This score is higher if the fixated locations along a scanpath have higher saliency as compared to rest of the input. Human response is expected to have a high score and a model scoring higher than the other will be considered better. The score is computed as $s = (1/N) \sum_{f_k \in T} S(f_k) - \mu_S$ in which the first term corresponds to the average value of N fixations f_k from an eye trajectory T. The second term μ_S is average value of the saliency map.

A method to evaluate saliency maps by comparing them to a benchmark map is proposed in [12]. A bi-directional comparison procedure compares only the salient areas of the source map with the corresponding areas of the target map. The source and the target are swapped while processing in the opposite direction. In each direction d the matching points m_d out of the total compared points N_d^s are counted. The similarity measure in one direction is taken as $\sigma_d = m_d/N_d^s$ which will give a 1 for identical saliency areas and 0 for a total mismatch. The optimistic similarity $\sigma^o = max(\sigma_1, \sigma_2)$ is considered as the measure of agreement between the two maps.

The model in [13] measures the improvements in selectivity gained by using top-down attention by counting the number of FOA hits and misses of traffic-relevant items like signal-boards and cars in video streams. The evaluation is feedback oriented as human viewers have to decide if the model has fixated on a correct location or not. For evaluation, a FOA is counted as a hit if at least half of the target object is within the FOA. An FoA is considered as a miss when a non-target is found while traffic-relevant and undetected targets still exist in the scene. The completeness is defined as the ratio of undetected targets that have been left in the image to detected targets. A high completeness score is considered as a measure of success.

The evaluation method for top-down attention in [14] measures performance of the search system using two metrics, namely, the average number of fixations per search, and the average search time in seconds. Less number of fixations before reaching the target is considered as indication of success. Shorter search time of course comes as a byproduct. The search model in [15] also uses the criteria of hit number to reach the target for evaluation of top-down search performance. They also use a metric of detection rate in which they measure if the target was fixated within the first 10 FOAs. The model presented in [16] performs evaluation of the top-attention by measuring the reaction time versus the number of items in a display. They also use number of attentional shifts before detection of target as a metric.

The assessment scheme in [17] has used percentage of erroneous fixations with respect to the total number of target elements as a measure of performance and percentage of erroneous fixations against quantity of distortions produced by different compression rates as a measure of robustness. In [18] the runtime of the model for a certain number of foci of attention is used as a measure of competence.

3 Analysis

In this section we analyze the exiting evaluation techniques in light of the questions mentioned in the first section. Answers to the said five questions will follow a hierarchical fashion because different answers to the first question will lead to branches each of which will require an independent answer to the second question. This branching will continue down to the fifth question. Figure 1(a) shows a tree built by possible answers to these questions. A parse through this tree from its root to one of the leaves will formulate the nature of the evaluation methodology. As answer to the first question regarding aspect of attention the current literature indicates two aspects namely, bottom-up (B) and top-down (T). As some methodologies can be used to evaluate either of the two pathways we can add an option of pathway-neutral (N) as an answer to the first question. For each of these attention aspects we can have two possible answers to the second question regarding aspect of evaluation, which are validation-test (V) and competence-test (C). Validation tests verify the correctness of a model's results by comparing them with benchmark results while competence tests measure the level of success or efficiency in context of a particular task, for example ability to cover target locations in restricted number of fixations.

The design of tests in each evaluation aspect will requires a specific type of output; hence the answer to the third question is a list of output-styles that can be produced by attention models. We include the commonly known styles of saliency map (M), fixation points (F), and scan path (P) in the current list. As each of these results can be viewed through different perspectives such as locations (L), saliency magnitudes (S), and order of sequence (O), therefore we assign these three answers to the fourth question.

Each of the above mentioned output perspectives has to go through a measurement methodology, which will convert the data of attention output into a set that could be used in comparison of models. The measurement methods used so far in the available evaluation schemes can be categorized into three groups, namely equivalence (E), feedback (D), and runtime of attention model (R). The equivalence methods compare the model's output with the benchmark results using some algorithms running on a computer while feedback methods use opinion of human subjects to measure the acceptability of the results. The underlined labels in figure 1(a) represent copies of the node represented by that label, for example $\underline{V}$ means that the hierarchy below V is repeated at the place.

There has to be one or more metrics for each methodology articulated by following a parse through the tree given in figure 1(a). Here we gather a set of commonly used metrics and arrange them into a small hierarchy in order to elucidate their utility. We organize the metrics into three categories namely *degree* (ε), *rate* (σ), and *similitude* (ψ). The *degree* metrics evaluate the performance of

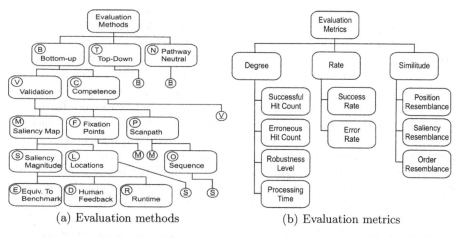

(a) Evaluation methods   (b) Evaluation metrics

Fig. 1. Hierarchichal organization of methods and metrics for evaluation of visual attention models

Table 1. Analysis of existing techniques for evaluation of attention models. Column headers: AA = attention aspect, EA = evaluation aspect, OF = output format, PR = perspective of results, MM = measurement method, MC = metric category. See section 3 for description of symbols given in columns 2 to 5. The subscripts in the 6th substitutes for ϕ in ε_ϕ^r and ε_ϕ^t. $\phi = N$ for noise, for rotations about angles of 90, 180 and 270 degrees we have $\phi = R90$, $\phi = R180$, and $\phi = R270$ respectively, $\phi = S$ for size of input image, $\phi = DR$ for distractors, $\phi = DN$ for distortion in input, $\phi = ST$ for search time, and $\phi = ET$ for exploration time.

Method	AA	EA	OF	PR	MM	MC	Metric
[1]	B	C	P	L	D	ε,σ	$\varepsilon^e,\ \varepsilon_N^r,\ \sigma^s$
[2]	B	V	P	L	E	ψ	ψ^a
[3]	N	V	P	L	D	σ	σ^e
[4]	B	V	F	L, O	D	ε	ε_ϕ^a
[6]	N	V	F	L, O	E	ψ	$\psi^a,\ \psi^o$
[7]	N	V	F	L, O	E	ψ	$\psi^a,\ \psi^o$
[8]	N	C	F	L	E	$\sigma,\ \varepsilon$	$\sigma^e,\ \psi^f,\ \varepsilon_{R90}^t,\ \varepsilon_{R180}^r,\ \varepsilon_{R270}^t$
[9]	N	V	P	L, O	E	ψ	$\psi^a,\ \psi^o$
[10]	B	V	M	L	E	ψ	ψ^a
[11]	B	C	M	S	E	ε	ε^s
[12]	B	V	M	L	E	$\psi,\ \varepsilon$	$\psi^a,\ \varepsilon_S^t$
[13]	T	C	F	L	D	ε	$\varepsilon^s,\ \varepsilon^e$
[14]	T	C	F	L	D	ε	$\varepsilon^e,\ \varepsilon_{ST}^t$
[15]	T	C	F	L	D, R	$\sigma,\ \varepsilon$	$\sigma^e,\ \varepsilon_{DR}^t$
[17]	B	C	F	L	D	σ,ε	$\sigma^s,\ \varepsilon_{DN}^r$
[18]	N	C	F	L	R	σ	ε_{ET}^t

a model on measures involving only one quantity like number of attempts before fixating on the target, the *rate* metrics consist of ratio between two quantities such as rate of successful fixations against total attempts, and the *similitude* metrics quantify the level of correspondence between model's output with some benchmark results. In the existing literature metrics belonging to the *degree* category include the measures like count of successful hits (ε^s), count of erroneous hits (ε^e), level of robustness against a given phenomenon (ε^r_ϕ), level of approval of results by human observers in context of a certain phenomenon (ε^a_ϕ), and processing time against some phenomenon (ε^t_ϕ) like time taken against number of distractors. Under the category of *rates* we can include error rate (σ^e), success rate (σ^s), and error rate depending upon some phenomenon (σ^e_ϕ) such as number of errors against noise in the input. In *similitude* category, the metrics of match between focused areas (ψ^a), resemblance of positions of fixations (ψ^f), correspondence of saliency magnitudes (ψ^s), and similarity between order of fixations (ψ^o) can be found in literature. Figure 1(b) demonstrates the hierarchy of these metrics in pictorial form. Table 1 summarizes the results of analysis of the existing evaluation techniques according to the above mentioned criteria. The techniques marked as N have performed evaluation of bottom-up attention but they are categorized as pathway-neutral keeping in view their extendibility to top-down attention.

4   Proposed Metrics and Methods

In this section we propose some metrics and methods to evaluate performance of attention models in which the outcome of the efficiency measure would remain between 0 and 1, 0 being the worst performance and 1 representing the best. We have concentrated on metrics that measure model performance against already known number and locations of targets because it is a better way to benchmark efficiency of models and compare them with each other.

4.1   Competence Tests

At first, we would like to introduce an evaluation method to measure the capability of a model to respond on a particular phenomenon, for example testing the capability to detect saliency with respect to a specific feature in specially designed synthetic images in which one or more objects are explicitly salient due to the examined feature. The measure of performance on that particular feature can be the ratio between the number embedded salient objects N_s and the number of objects found by the model, N_f. As this metric is meant for testing ability of a model to fulfill a specific purpose therefore the model must find the given objects within N_s fixations, i.e., we are judging whether the model is suitable for the purpose or not. Hence, for a purpose ϕ, the measure of success σ^s_ϕ will be computed as

$$\sigma^s_\phi = \frac{N_f}{N_s} \quad f \in \{1, ..., N_s\}$$

As $N_f \leq N_s$ is true in all cases due to the counting method described above, hence $0 \leq \sigma^s_\phi \leq 1$. A slightly modified version of this metric can be used for time critical systems in which the targets have to be detected in a particular period of time. In this case the system will be allowed to fixate for a time t and the number of attended locations will be counted as N_t. If, out of these attended locations, N_s^t locations are among the targets then, keeping in view that $N_s^t \leq N_t$ is always true, the success rate $\sigma_\phi^s(t)$ of the system will be calculated as

$$\sigma_\phi^s(t) = \frac{N_s^t}{N_t}$$

The generalized form of the above mentioned metrics is to judge a model's capability to cover the already known number of targets without specification of a particular phenomenon, as done by some of the existing techniques. Such metric is pathway-neutral because the targets could be considered salient in bottom-up context and also as pre-defined objects (or locations) that should be attended in a top-down search. Having N_s salient objects in a given-scene a model will be allowed to keep on fixating for N_a number of times until all required locations are covered, hence $N_a \geq N_s$ is always true. Practically it is possible that a model may not be able to reach some targets at all. Therefore, in order to avoid running of a system for an indefinite period of time we propose to impose a maximum limit for N_a. In most of that cases this limit could be set to N_s^2 while for $N_s \leq 2$ it could be set to a constant value such as 10 or 15. Now the generalized detection rate σ^d of the model can be defined as

$$\sigma^d = \frac{N_s}{N_a}$$

An efficient model will yield $\sigma^d = 1$ whereas detection rates close to zero will show inefficiency.

For quantifying the general error rate of a model we propose a simple method in which the fixations falling outside the target locations will be counted as erroneous fixations N_e. Its ratio to the fixations taken to cover all targets N_a gives the error rate σ^e as

$$\sigma^e = \frac{N_e}{N_a}$$

We introduce another metric using which the explorative capabilities of models will be measured. This measure will be the inverse of the tendency of a model to repeat fixations on already attended locations. If N_D distinct targets from the total N_s locations are covered by a model in N_a fixations then the degree of exploration capability ε^x will be computed as

$$\varepsilon^x = 1 - \frac{N_a - N_D}{N_a}$$

A model that does not repeat fixation on any of the attended locations before covering all salient locations will yield a value of 1 for ε^x while lower values will be obtained for model that tends to revisit already attended locations.

4.2   Validation Tests

For location sensitive evaluation, we propose to measure average distance between fixations performed by the two models under consideration. The model

m under evaluation will be allowed to fixate for N_a times while the benchmark dataset consists of N_s salient locations. The evaluation method will pick the fixated locations, F_i^b, from the benchmark dataset one by one and find the distance of F_i^b from the corresponding fixation F_j^m by the model m. Having such distances for n locations the performance measure ψ^f of the model m with respect to the benchmark model in terms of similarity of fixation positions will be computed as

$$\psi^f = 1 - \frac{\sum_{i=1}^{N_s} \Delta \left(F_i^b, \aleph(F_i^b, F_j^m \ \forall \ 1 \leq j \leq N_a) \right)}{N_s \Delta^{max}}$$

where $\Delta(.,.)$ computes the spatial distance between two given locations. It can be extended to distance between two locations in space when carrying out experiments on active vision. $\aleph(.,.)$ picks the fixation F_j^m from the set of N_a fixations by the model m that corresponds to the currently picked benchmark fixation F_i^b. To be designated as a corresponding fixation, the F_j^m should be on the same location as F_i^b, cover the object pointed by F_i^b fully or partially, or be on an adjacent object having feature similarity to F_i^b. When no corresponding F_j^m is found for F_i^b, $\Delta(.,.)$ returns Δ^{max} to reduce the magnitude of ψ^f. The F_j^m once found corresponding to some F_i^b is excluded from further processing. Δ^{max} is the maximum distance that can be involved during the current experiment. For a rectangular image with length L and width W, we suggest to compute Δ^{max} as

$$\Delta^{max} = \sqrt{L^2 + W^2}$$

The value of ψ^f will be close to 1 for a good equivalence between the given model and the benchmark while lower values will reflect a poor match.

For sequence sensitive evaluation, the difference of sequence number between the currently picked benchmark location F_i^b and the corresponding F_j^m will be determined. Having data of such differences for N_s fixations, the sequence disparity measure ψ^o will be computed as

$$\psi^o = 1 - \frac{\sum_{i=1}^{N_s} \Delta^o \left(F_i^b, \aleph(F_i^b, F_j^m \ \forall \ 1 \leq j \leq N_a) \right)}{N_s N_a}$$

where $\Delta^o(.,.)$ calculates the absolute difference between the sequence order numbers of the given F_i^b and its corresponding F_j^m. When no corresponding fixation is found for F_i^b, $\Delta^o(.,.)$ returns N_a to depreciate the overall value of ψ^o. For a perfect match of sequence ψ^o will gain a value of 1 while poor similarity will be represented by lower values down to zero.

5   A Sample Implementation

The metrics proposed in section 4 were applied to three different attention models proposed in [1], [3], and [19] respectively. It may be noted that the objective

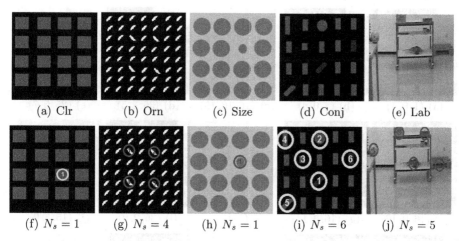

(a) Clr   (b) Orn   (c) Size   (d) Conj   (e) Lab

(f) $N_s = 1$ (g) $N_s = 4$ (h) $N_s = 1$ (i) $N_s = 6$ (j) $N_s = 5$

Fig. 2. (a) to (e) Samples from input dataset used in evaluation experiments. Captions under the images represent the code used as reference for respective image. (f) to (j) Salient locations marked by human observers.

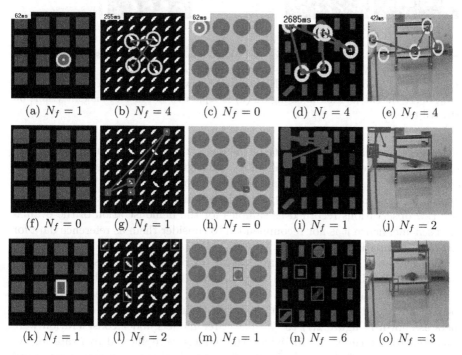

(a) $N_f = 1$ (b) $N_f = 4$ (c) $N_f = 0$ (d) $N_f = 4$ (e) $N_f = 4$

(f) $N_f = 0$ (g) $N_f = 1$ (h) $N_f = 0$ (i) $N_f = 1$ (j) $N_f = 2$

(k) $N_f = 1$ (l) $N_f = 2$ (m) $N_f = 1$ (n) $N_f = 6$ (o) $N_f = 3$

Fig. 3. (Top row) N_s fixations performed by Itti's model [1] on images shown in figure 2. (Middle row) N_s fixations by E-Saliency model [3]. (Bottom row) N_s fixations by Region-Based model [19]. The fixation by this model in the left most column is repainted in order to improve its visibility on the red region.

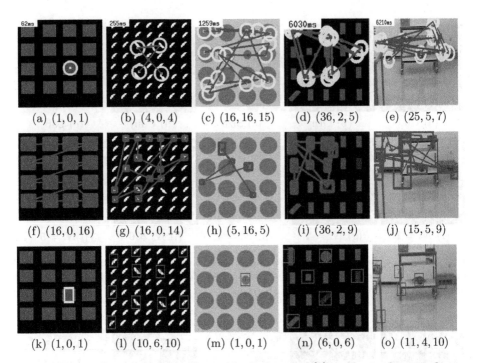

Fig. 4. (Top row) N_a fixations performed by Itti's model [1] on images shown in figure 2. (Middle row) N_a fixations by E-Saliency model [3]. (Bottom row) N_a fixations by Region-Based model [19]. The triplets written below each image represent the values (N_a, N_e, N_D). The fixation in subfigure (k) is repainted in order to make it visible on the red region.

here is not to measure competence of the models rather to demonstrate working of the proposed evaluation methodology. Strength of each model under discussion may lie in specific areas and the used input samples do not necessarily represent those particular aspects. It is important to declare that these results are raw observations recorded as samples for the demonstration of the evaluation scheme, hence it is not recommended to consider them as reference data for actual benchmarking purposes.

The five input samples in figure 2(a) to (e) represent the test data used in evaluation experiments. For these experiments the N_s salient objects were determined by human observers by marking those object that are most likely to be focus of attention. They are marked with their most probable sequence numbers as shown in figure 2(f) to (j). The three models under consideration were executed to determine N_f for which the models were allowed to fixate for N_s times for each input. Similarly N_a was found by letting the systems fixate until all of the N_s locations were covered or the maximum limit of N_a was reached. Figure 3 shows the fixations by the three models while finding out N_f and figure 4 shows the fixations for N_a times. The computed values of σ_ϕ^s, σ^d, ε^x, ψ^f, and ψ^o are plotted in figure 5.

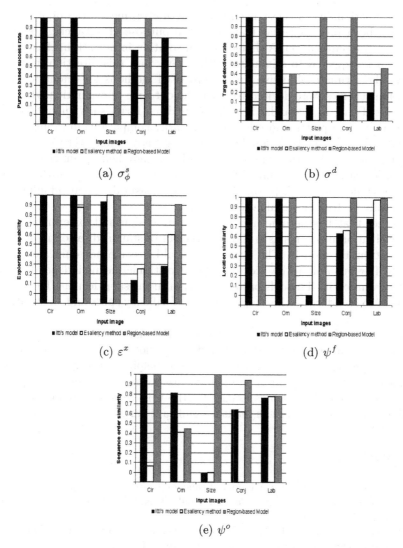

Fig. 5. (a) to (c): Plot of values obtained for success rate σ_ϕ^s, target detection rate σ^d, and degree of exploration capability ε^x for the models of [1], [3], and [19]. (d) and (e): Plot of values obtained for fixation position match ψ^f and sequence order match ψ^o against the human response shown in figure 2.

6   Conclusion

Standardization of yardsticks and methods to obtain comparable quantities is very important for truthful growth in every field of science. Keeping in view the deficiency of such standards for evaluation of attention models we have performed a survey of the existing techniques and tools for this purpose. After analyzing the utility and functionality of these metrics and methods some metrics of objective nature have been proposed that may serve the purpose of advancing towards the required standardization. Although a final solution in not attained but this contribution is expected to stimulate further work into this direction in order to achieve the ultimate objective. In the proposed methods for evaluation the computation schemes were devised to make the outcome of evaluation within a unit amount. Being in a standardized or normalized form, such quantities are easily scalable when plotting graphs and also easily comparable in tabular representations. This can be considered as a significant advantage over the existing metrics. The proposed methods for extracting basic ingredient quantities for the metrics can be carried out by human feedback as well as through algorithmic approach. These methods are kept simple and easily measurable to avoid influence of non-relevant factors into evaluation process.

One aspect that has become noticeable during design and experimentation of these evaluation mechanisms is the requirement of standardization in output format of attention models. The evaluation would become not only more objective and fair but able to be processed by algorithmic means independent of human intervention if all models produced output in the same format. There is a special need to standardize the shape and size of saliency spots and the fixation windows.

The range of values that represent the saliency magnitudes should also be standardized so that the comparisons of models may not get disrupted due to different magnitude representations. It would also be helpful in justified evaluation if the images used by the models had standardized dimensions and bit-depths. This homogenization is needed especially for the time sensitive comparisons in which difference of image size can make a significant variation in response time.

References

1. Itti, L., Koch, U., Niebur, E.: A model of saliency-based visual attention for rapid scene analysis. Transactions on Pattern Analysis and Machine Intelligence 20, 1254–1259 (1998)
2. Peters, R.J., Iyer, A., Itti, L., Koch, C.: Components of bottom-up gaze allocation in natural images. Vision Research 45, 2397–2416 (2005)
3. Avraham, T., Lindenbaum, M.: Esaliency - a stochastic attention model incorporating similarity information and knowledge-based preferences. In: WRUPKV-ECCV 2006, Graz, ECCV 2006 (2006)
4. Baltazar, J., Pinho, P., Pereira, F.: Visual attention driven image to video transmoding. In: Picture Coding Symposium (PCS 2006), Beijing, China (2006)
5. Chen, L., Xie, X., Fan, X., Ma, W., Zhang, H., Zhou, H.: A visual attention model for adapting images on small displays. ACM Multimedia Systems Journal 9, 353–364 (2003)
6. Privitera, C.M., Stark, L.W.: Algorithms for defining visual regions-of-interest: Comparison with eye fixations. Transactions on Pattern Analysis and Machine Intelligence 9, 970–982 (2000)

7. Clauss, M., Bayerl, P., Neumann, H.: A statistical measure for evaluating regions-of-interest based attention algorithms. In: Rasmussen, C.E., Bülthoff, H.H., Schölkopf, B., Giese, M.A. (eds.) DAGM 2004. LNCS, vol. 3175, pp. 383–390. Springer, Heidelberg (2004)

8. Draper, B.A., Lionelle, A.: Evaluation of selective attention under similarity transforms. In: WAPCV 2003 (2003)

9. Marmitt, G., Duchowski, A.T.: Modeling visual attention in vr:measuring the accuracy of predicted scanpaths. In: EUROGRAPHICS 2002 (2002)

10. Meur, O.L., Callet, P.L., Barba, D., Thoreau, D.: A coherent computational approach to model bottom-up visual attention. Transactions on Pattern Analysis and Machine Intelligence 28, 802–817 (2006)

11. Hügli, H., Jost, T., Ouerhani, N.: Model performance for visual attention in real 3D color scenes. In: Mira, J., Álvarez, J.R. (eds.) IWINAC 2005. LNCS, vol. 3562, pp. 469–478. Springer, Heidelberg (2005)

12. Aziz, M.Z., Mertsching, B.: Fast and robust generation of feature maps for region-based visual attention. Transactions on Image Processing 17, 633–644 (2008)

13. Michalke, T., Gepperth, A., Schneider, M., Fritsch, J., Goerick, C.: Towards a human-like vision system for resource-constrained intelligent cars. In: ICVS 2007, Bielefeld University eCollections, Germany, pp. 264–275 (2004)

14. Hawes, N., Wyatt, J.: Towards context-sensitive visual attention. In: Second International Cognitive Vision Workshop (ICVW 2006) (2006)

15. Frintrop, S., Backer, G., Rome, E.: Goal-directed search with a top-down modulated computational attention system. In: Kropatsch, W.G., Sablatnig, R., Hanbury, A. (eds.) DAGM 2005. LNCS, vol. 3663, pp. 117–124. Springer, Heidelberg (2005)

16. Navalpakkam, V., Itti, L.: Modeling the influence of task on attention. Vision Research, 205–231 (2005)

17. Aziz, M.Z., Mertsching, B.: An attentional approach for perceptual grouping of spatially distributed patterns. In: Hamprecht, F.A., Schnörr, C., Jähne, B. (eds.) DAGM 2007. LNCS, vol. 4713, pp. 345–354. Springer, Heidelberg (2007)

18. Aziz, M.Z., Mertsching, B.: Pop-out and IOR in static scenes with region based visual attention. In: WCAA-ICVS 2007, Bielefeld - Germany, Bielefeld University eCollections (2007)

19. Aziz, M.Z., Mertsching, B.: Color saliency and inhibition using static and dynamic scenes in region based visual attention. In: Paletta, L., Rome, E. (eds.) WAPCV 2007. LNCS (LNAI), vol. 4840, pp. 234–250. Springer, Heidelberg (2007)

Comparing Learning Attention Control in Perceptual and Decision Space

Maryam S. Mirian[1], Majid Nili Ahmadabadi[1,2], Babak N. Araabi[1,2], and Ronald R. Siegwart[3]

[1] Control and Intelligent Processing Centre of Excellence
Dept. of Electrical and Computer Eng, University of Tehran
[2] School of Cognitive Science, IPM, Tehran, Iran
{mmirian,mnili,araabi}@ut.ac.ir
[3] ASL, ETHZ, Switzerland
rsiegwart@ethz.ch

Abstract. The first question answered in this paper is whether or not learning attention control in the decision space is feasible and how to develop an online as well as interactive learning approach for such control in this space, in case of feasibility. Here, decision space is formed by the decision vector of the agents each has allowed to dynamically observe just a subset of all available sensors. Attention control in this new space means active and dynamic selection of these decision agents to contribute in making final decision. The second debate is verifying the advantages of attention control in decision space over that in perceptual space. According to the tight coupling of attention control and motor action selection, in order to answer above mentioned questions, attention control and motor action selection are formulated in a unified optimization problem and reinforcement learning is utilized to solve it. In addition to the theoretic comparison of learning attention control in perceptual and decision space in terms of computational complexity, two proposed approaches are tested on a simple traffic sign recognition task.

Keywords: Attention Control, Learning, Multi-modal perceptual space, Decision fusion, Mixture of Experts, Soft Decision.

1 Introduction

Basically, attention control can be assumed as an active intelligent filter which trims down the dimension of the huge input sensory space and prevents reaching it entirely to the further processing units. In other words, it is a must for an agent to purposefully reduce the computational burden of sensory input processing before performing any cognitive task; such as object recognition or scene interpretation.

The great significance of attention control is in fact because of these requirements: reduction of probable confusion among multiple dimensions of the perceptual space, faster response and dealing with dynamicity of perceptual space. The mentioned dynamics is in sense of reliability and accuracy of multiple sensors or processing elements. These requirements in face of limited processing power necessitate a dynamic

L. Paletta and J.K. Tsotsos (Eds.): WAPCV 2008, LNAI 5395, pp. 242–256, 2009.
© Springer-Verlag Berlin Heidelberg 2009

attention control strategy rather than designing just a simple sensor selection algorithm. The tight coupling of attention control and motor action selection in a sequential decision making is another concern which makes the problem even more challenging. There are not enough works done in the field of learning attention while learning the desired behavior. It means attention control strategy is task-dependent. As a result, we couple motor actions with those that are performed solely for change of attention focus. The later ones are called perceptual actions and include those are mental only – like giving more weights to color in comparison to shape for example- and the actions that involve control of physical sensors –such as saccadic movements. We call selection of pure motor actions decision making.

It is clear that information bottleneck gives meaning to attention control however; here we raise this question that what type of information should be attentively processed? In other words, we are interested to know if attention control is restricted to active sensor selection or there is another information space where attention control can be learnt more effectively or robustly. In this paper, we chose decision space – or more accurately the probability vector of selecting actions- as a candidate information space to apply attention control in it and compare the results with those we attain in the sensory space. To perform the mentioned comparison, we model attention control as an optimization problem and choose reinforcement learning for solving this problem. The reason behind such a choice is to provide the potential for interactively solving the problem when the agent is acting in its world. By doing so, the agent learns the attention control strategy in concert with learning its task in the framework of expected reward maximization.

In this paper, we first review the related works on learning attention control. After that, two proposed approaches are described in details. Then, we will express the testbed and the results taken. Finally a comprehensive discussion, conclusions and future works are given.

2 Related Works

Surely, we implicitly know what we mean by attention. But, a psychological definition may be a good starting point: focusing mind in a clear manner on one of many subjects or objects that may simultaneously stimulates the mind [1]. Adopting engineering perspective, it can be considered as a filtering process which trims down the input sensory space to help us focusing on some thing which is more valuable to be processed, i.e., worth-focusing. Let's look at the attention problem from action perspective and this means using active perception instead of processing the entire sensory space. This is the viewpoint we have adopted and tried to realize it through learning. In this section, the review of related works is done with more focus on learning aspects of attention. Unfortunately, there are a few researches on learning and formation of attention control; rather they are mostly related to the attention modeling. [2] presents an RL[1] based approach in which visual, cognitive and motor processes are integrated to help an agent learn how to move its eyes in order to generate an efficient behavior of a human expert while reading. Using two spatial and temporal

---

[1] Reinforcement Learning.

modeling parameters (fixation location of eyes as well as their fixation time) the optimal behavior is learned. In [3] a framework for attention control is presented which performs actively in high level cognitive tasks. It contains three phases: the first phase is learning attention control as in active perception. Then in the second phase it extracts those concepts learned previously and finally using mirror neurons it abstracts the learned knowledge to some higher level concepts. Continuing this work is one of our main motivations, but we are focused here on learning in the decision space rather than in perceptual space. In [4] attention control is applied in object recognition task but in a limited image database. The main idea is using information theoretic measures to find discriminative regions of the image in a general to specific manner. In [5], as a continuation of [4], a 3-step-architecture is presented which firstly extracts attention center according to information theoretic saliency measures. Then, by searching in pre-specified areas found from first step decides whether the object is available in the image and finally a shift for attention will be suggested. The final step is done using Q-Learning with the goal of finding the best perceptual action according to the search task. This research is related to our work because it also couples decision making and attention control and uses reinforcement based learning approach. In [6] two approaches for attention control are presented in a robotic platform with neck, eyes and arms. The first approach is a simple feed forward method uses back-propagation learning algorithm while the second uses reinforcement learning and a finite state machine for state space representation. Their results confirm that the second approach generates better performance in terms of finding previously observed objects even with fewer movements in head and neck and also in attention center shift. In [7] some approaches based on hidden states in reinforcement learning are proposed for active perception in human gesture recognition. This work proposes some solutions for perceptual aliasing. This problem is realized when there is a many to many correspondence among environment's state and agent's state. In such a situation, the agent's decision making has ambiguity and in order to reduce it, the agents decide to perform perceptual actions. This problem can be handled by merging similar (from utility perspective) states or splitting one state due to non-homogeneity in utility measure. The approaches for merging / splitting states presented in [7] are called Utile Distinction Memory and Perceptual Distinction Approach. Moreover, in order to handle the problem of requiring more than one shot observation, an approach called Nearest Sequence Matching is proposed which uses a chain of recent observations (state / action) to declare current state. The results show that by learning, they can find more informative set of features to attend for gesture recognition rather than just selecting them in a pre-specified manner. Unfortunately, it is mentioned that the computation load of these approaches are very high and can be problematic in real complex applications. In papers reviewed till now, the control policy was spatial. In [8] some biological evidences are presented which show that attention can also be directed to particular visual features, such as a color, orientation or a direction of motion. They showed effects of shifting attention between feature dimensions, rather than specific values of a given feature. In one condition the monkey was required to attend to the orientation of a stimulus in a distant location. In a second condition it was required to attend to the color of an un-oriented stimulus in the distant location. Finally, inspired from *Mirror Neuron* idea in [9], there is an indirect biological support for the action-based representation in the decision space as what we proposed in this paper. So, it can be assumed that for each stimulus in perceptual space, there is a corresponding action-based representation in the decision space and we have proposed two approaches for learning attention control in both spaces. Furthermore, according to the discussion presented by Rizzolatti in [10], there is a close relation between attention processes and motor planning processes. In fact, as claimed in their theory, there is a strict link between covert orienting of attention and programming explicit ocular movements.

3 Our Approach

Two approaches proposed here are based on these main concepts: Virtual Sensors and Decision Agents. Before going further into details, we define *virtual sensor* and *decision agent*. A *virtual sensor* is a processing element that gets the sensory information and extracts some high level features. A physical sensor can be regarded as a virtual sensor with the identity information processing function. According to this definition, attention control mechanism controls the physical sensors as well as the virtual ones, see Fig. 2.

A *decision agent* is a processing unit that resides inside the main agent and looks at the world through a set of virtual sensors. Its output is a probability vector. Element i of that vector is the suggested probability of selecting action i by that decision agent. Note that each action can be a pure motor action, a perceptual one or a combination of both. See Fig. 3.

As mentioned before, we have taken some primary steps to resolve the main problem of proposing a general framework for learning attention control in a dynamic and multi-modal perceptual space. Since, attention control and decision making are very closely correlated problems, we employ attention control alongside of decision making once in a high-dimensional perceptual space and once in a decision space. Therefore, in this paper, two models are proposed for a sequential, multi-step learning in each high dimensional space and the advantages and disadvantages are verified.

To summarize, in sensory space, based on the agent's current state, it learns which virtual sensors to look at in the next step in order to make the most beneficial decision, see Fig. 2. In that figure, the agent is at state S and has a set of action pairs each composed of a motor actions and a perceptual one; i.e. $A=\{(a_P, a_M)\}$ where A is the agent's action set. In other words, the agent takes a perceptual action (a_P) to select a virtual sensor and a motor action (a_M) to affect its environment.

Similarly, in decision space, the agent tries to find those decision agents –or local experts as such entities are named in multi-agent domain- to consult with to find the best decision, see Fig. 3. Again, in this scenario, the agent employs its perceptual action to select a decision agent. Note that any attentive selection –either selection of a virtual sensor or choosing a decision agent- involves processing the related sensory information.

In addition, it is worth mentioning that the selection strategy is sequential. It means that a selection is done after the selected entities are processed. It is also important to note that, similar to any motor action, each virtual sensor selection (and its processing) or expert consultation has a cost and the agent needs to minimize the total cost. The associated cost is related to the complexity of each virtual sensor or decision agent.

As Fig. 2 shows, learning attention control in the sensory space is straightforward. The agent tries to select (or in fact to attend to) those more relevant virtual sensors to the task at hand. It is done implicitly by learning a mapping between the agent's state and its optimum action in that state.

Learning attention control in the decision space looks more complex however; it is a new approach to the complicate problem of attention control. The approach benefits many interesting aspects of distributed and multi-agent systems as the agent's mind is composed of local decision makers each looking at a portion of sensory information. These local decision makers (when trained) form our local experts and the final decision of the agent can be shaped through a *mixture of experts* strategy.

Although the real world can be modeled by a MDP [11] from an absolute agent's point of view, our agent is a partial observer. So, we need to propose a POMDP approach. But to keep the problem manageable at this stage, we considered one coupled optimization problem in MDP framework. The Markov decision process provides the general framework to outline sequential attention for optimal decision making. A MDP is defined by a 4-tuple (*States, A, δ, R*) with state set *States*, action set *A*, probabilistic transition function δ and reward function R. In each transition, the agent receives reward from a critic according to $R : S \times A \to R$, $Rt \in R$. The agent must act to maximize the utility $Q(s, a)$. The decision process in sequential attention control is determined by the sequence of choices on perceptual actions - either in sensory or decision space- at specific states, see Fig. 1.

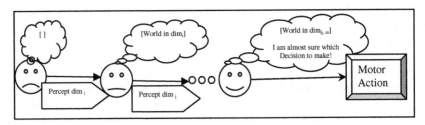

Fig. 1. A simple view of sequential perceptual state change

Fig. 1 simply shows the sequential change of agent's mental state due to performing multiple perceptual actions. At first, the agent's state is null, i.e. it knows nothing about the world's state. After a while, it decides to percept dim_i and its state changes accordingly. This continues until it can specifically decide which motor action is the most suitable to be performed. The following sections explain both learning models former in sensory space and latter in decision space.

3.1 Approach 1- Learning Attention control in Perceptual Space: Attentive Sensor Selection

In this approach, we want the agent to learn which features to attend in a state in order to gain maximum reward or in fact can perform the task as efficiently as possible from the critic's perspective, See Fig. 2.

Assume that the agent is allowed to use maximum m physical sensors to percept the environment and based on this information should perform one best *action* among

k available actions consisting of both perceptual and motor actions. Each physical sensor is equipped with a set of processing layers, let's say n. As mentioned before, we can assume each physical sensor plus its processing layer as a virtual sensor. The agent can either turn on all sensors at once which is very computationally expensive, time-consuming and maybe redundant or it can try to build up its percept based on a subset of its whole sensors; here, those it has found more rewarding. This can be thought as a very rough definition of agent's attention control problem. When a learning episode starts, the agent should decide whether to perform more perceptual actions to reduce ambiguity in its perception or just perform a motor action and terminate the episode. In this setting, action and state sets (A and S respectively) are defined as:

$$A = \{perceptual\_action, null\} \text{ x } \{motor\_action, null\} \tag{1}$$

$$S = \{s = (o_1, o_2, ..., o_m) : o_i = f_j(sensor_i) \} \quad i = 1,..., m \qquad j = 1,..., n \tag{2}$$

Where

$$f_j(sensor_i) \in \{v_1, v_2, ..., v_f, null\} \tag{3}$$

the output value of each sensor processing takes maximum f_i+1 values for sensor i including *null* when that sensor is not attended. For example, if we have a virtual sensor for temperature with three fuzzy labels, a two-valued-color and a two-valued-shape, S is:

$S=$ {Hot, Cold, null} x {Red, Blue, null} x {Circle, Rectangle, null}. Note that a learning episode start from the null state and after a number of perceptions or after a time, when a motor action is performed, the current episode will end. Performing perceptual actions have different constant costs. This cost is a function of power consumption of the sensor and the associated processing time of its processing function. Also, when a correct motor action is performed a positive value is assigned to it. This is the common strategy of *Reward Function* of the MDP frameworks used in both approaches. Fig. 2 is a schematic view of the proposed decision making strategy coupled with attention problem (from sensor selection perspective) using RL as a learning

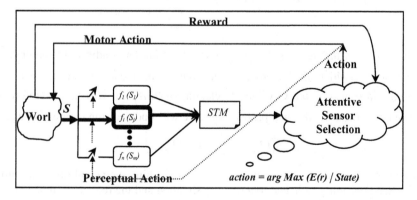

Fig. 2. Schematic view of attentive sensor selection method

method. In Fig. 2, f_1, f_2,..., f_n are processing functions like: dominant color finding, color segmentation, shape detection, straight line extraction, template matching on vision sensor and so on. STM is short term memory and here keeps the required present and past observations. The agent's state is in fact kept in STM.

3.2 Approach 2- Learning Attention Control in Decision Space: Attentive Decision Fusion

In this section, a general method for learning attention control is proposed in the decision space, see Fig. 3. Here, one simple implication from the decision space is proposed.

Again assume we have m sensors each observed by a tiny agent. These tiny agents are in fact our *local decision makers*. When they learned the decision making task individually in their own partial sensory space (and the learning is saturated), they start to propose their decisions (if the fuser asked them) and based on their nongreedy opinions, the agent should make the best decision which is actually performing one *action* among k available actions. The agent can either consider decisions made by every local expert, which is not a reasonable policy, or it can learn to build up its decision profile based on a subset of the whole decision set and on a need basis. After this introduction, let's define the decision space:

Decision sub-space is a sub-space formed by Boltzman probabilities of selecting each motor action$_j$ on the condition of state$_{S_i}$ (as i-th sensor concerns) when the learning by agent$_i$ is finished.

It means for each partial observation done by each tiny agent, there is one selection probability for a motor action. This definition named "decision template" is similarly introduced in [12]. Putting these templates together we will find a decision profile. It is noticeable that instead of using greedy decisions of each agent (their hard decisions) we used their soft decisions in order not to miss any probably helpful information. The mathematical definition of this subspace is expressed here:

$$D_{j|O_i} = P(action_j \mid state = O_{S_i}) = \frac{e^{Q(state_{S_i}, action_j)}}{\sum_k e^{Q(state_{S_i}, action_k)}} \tag{4}$$

in which $D_{j|O_i}$ is the agent$_i$'s decision to select action$_j$ on condition to the environment state O_i (which is the environment state from agent$_i$'s point of view) and $Q(state_{S_i}, action_j)$ is the Q-value of selecting $action_j$ in $state_{S_i}$. Therefore, by concatenating these conditional probabilities, we will find decision template of agent$_i$:

$$D_{O_i} = \left[D_{1|O_i} \mid D_{2|O_i} \mid ...D_{M|O_i} \right] \tag{5}$$

in which M is number of motor actions. The reason behind such conditional definition is that each decision is attached to a specific situation and the real environmental state is the link of the local or partial states (observed by each tiny agent). As in *Attentive*

Sensor Selection, when a learning episode starts, the agent should decide whether to perform more perceptual actions (consult more experts) to find a more descriptive state or just perform a motor action and terminate the episode. Note that a learning episode start from the null state and after a number of perceptions or after a time, when a motor action is performed, the current episode will end. Performing perceptual actions (consultation with experts) have different constant costs. Also, when a correct motor action is performed a positive value is assigned. In this setting, action and state sets (*A* and *S* respectively) are defined as:

$$A = \{perceptual\_action,\ null\} \times \{motor\_action,\ null\} \tag{6}$$

$$S = \{s = (D_{O1}\ |\ null), (D_{O2}\ |\ null), ..., (D_{Om}\ |\ null)\} \tag{7}$$

Fig. 3 shows the learning strategy for decision making coupled with learning attention control in the decision space.

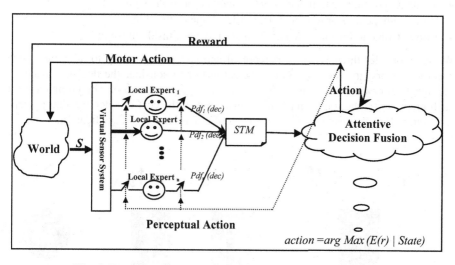

Fig. 3. The schematic view of Attentive Decision Fusion method

4   Testbed, Evaluation Measures and Simulation Results

In this section, first we introduce our testbed. Then the evaluation measures for comparing these two proposed learning strategies are defined. Finally the simulation results are given and analyzed.

4.1   Testbed

As a decision making problem, a simple cognitive task of Traffic Sign Classification is considered: "A*t the beginning of each episode, a single sign is shown to the agent. Using Attentive Sensor Selection or Attentive Decision Fusion it should decide which action to perform to minimize the total cost (of processing a feature or consulting a decision agent)*". There is a one to one correspondence between the signs and motor actions to perform. This is obviously a simple classification task which may be re-solved with no attention control policy. But, there are some reasons for selecting such testbed to test our basic ideas:

- Without losing generality, any real cognitive application can be considered as a classification problem with a vast number of classes and different input data and it has the potential of extension to more complex tasks.
- This is a primary step of our ongoing research and we need to gradually test the ideas and make sure if they work. Therefore, the complexity of task should be kept small enough in order not to dominate the learning strategy.
- It is surely required in any real autonomous vehicle driving / assistant application which maybe a very good testbed for this research according to the great need to attention control in such applications.

There are three virtual sensors for the agent to percept the environment:

- Virtual Color Sensor to detect the dominant color of the sign
- Virtual Shape Sensor to detect the border shape of the sign
- Virtual Content Sensor to detect the text or symbol inside the sign.

We can consider three types of perceptual actions corresponding to attending these specific sensors (in Attentive Sensor Selection) or to consider the decision made by the agent observes these sensors (in Attentive Decision Fusion). The complexity of each processing function is implicitly considered in the cost of selecting that percep-tual action. Fig. 4 shows the selected subset of traffic signs for classification.

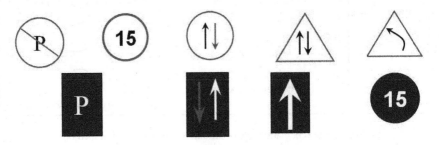

Fig. 4. Selected Traffic Signs for Recognition

According to the selected signs, we can define:

- C = Colors detected by Virtual Color Sensor = {Blue, Red}
- S = Shapes detected by Virtual Shape Sensor = {□, ○, △}
- CN = Contents detected by Virtual Content Sensor {P, 15, ↻, ↑, ↑↓}.

4.2   Evaluation Measures and Simulation Results

There are two sets of measures for evaluation of the proposed approaches. The first set which is tightly coupled to reward function design is accumulative reward and recognition rate. The second set contains secondary measures to evaluate our approaches: perceptual steps taken after learning and required number of episodes to complete the learning.

Approach 1- Learning Attention control in Perceptual Space: Attentive Sensor Selection

In order to show the effectiveness of the first approach, we compare it with the case where there is no attention control and the agent can utilize all its sensors at once. The results are shown in Table 1.

Table 1. Results of Simulating Approach 1 (Attentive Sensor Selection)

Measures	With Attention Control (Attentive Sensor Selection)	Without Attention Control
Recognition Rate after learning	100%	100%
perceptual steps taken	2.1	3
Average Reward gained during learning	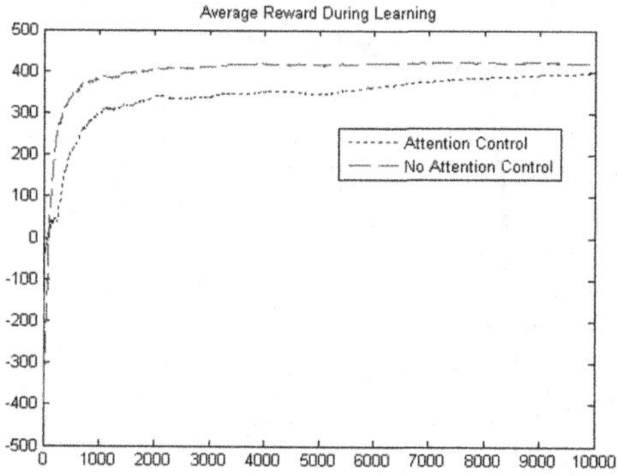	

Fig. 5. The accumulative reward during learning in perceptual space

The results justify that if we have enough time and processing power, there is no need to control the attention and the agent can learn the task even more quickly as its state space is three times smaller. However, when the attention control is necessary, *Attentive Sensor Selection* can gain perfect recognition rate while taking smaller number of perceptual steps; which means faster response and consuming less processing power.

In order to evaluate the amount of computational efficiency found by using the first approach, two other sets of results are also generated:

o Learning the task in uni-modular spaces
o Learning the task in bi-modular spaces (Color + Shape, Shape + Content and Color + Content): This is when the agent has pair of fixed sensors to percept the environment and selects its motor action accordingly.

Table 2 shows the recognition rate of the mentioned cases as well as the average reward of Attentive Sensor Selection vs. fixed bi-modular selection. The results clearly confirm that using *Attentive Sensor Selection* for attention control in the input sensory space can significantly enhance both the accumulative reward and also the recognition rate (a direct measure of success in decision-making). This is because, the agent autonomously and efficiently selects best pair of sensors to attend according to the state situated in, or maybe in some cases it pays to attend to all available sources to find the most rewarding decision.

Table 2. Results of Simulating Approach 1 (comparing with fixed selection in sub-modalities

Learning in Uni-modular Space		Learning in bi-modular Space	
Color	20%	Color + Shape	46%
Shape	30%	Shape + Content	80%
Content	50%	Color + Content	88%

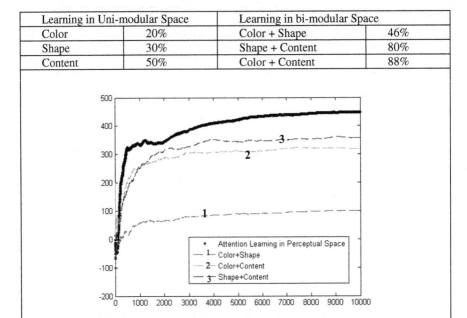

Fig. 6. The accumulative reward of Attentive Sensor Selection vs. fixed bi-modular sensor selection

Approach 2- Learning Attention Control in Decision Space: Attentive Decision Fusion

The effectiveness of the second approach (*Attentive Decision Fusion*) is shown in comparison with the first approach (*Attentive Sensor Selection*). Learning in decision space starts with learnt pre-knowledge of each decision agent. It means, in the first step the decision agents learn the task in a parallel manner. Then, each proposes a decision vector to the main agent. The main agent uses the Max operator and selects the action with the highest probability value. All decision agents update their knowledge knowing the selected action and received reward. The agent starts learning attention control in decision space when the first step is finished. Note that, to have a fair comparison with attention control in sensor space, the learning cost of the first step is added to the cost of attention control in the decision space. The results show that the agent has learned attention control in decision space however; learning attention control in decision space is slower than learning it in the sensory space. Moreover, the number of perceptual steps taken in decision space is larger than that in the perceptual space. A detailed comparison is given in the next section.

Table 3. Results of comparing two approaches

	Attentive Sensor Selection	Attentive Decision Fusion
Recognition Rate(Test)	100%	100%
Perceptual Steps (Test)	2.1	2.8
Required episodes[2]	1000	1900
Average Reward (Learning)	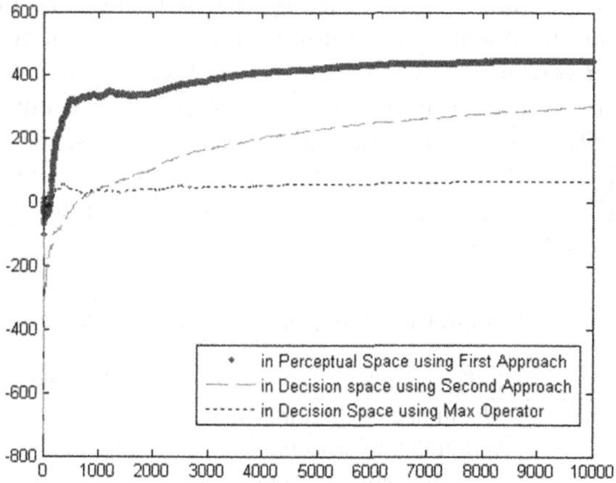	

Fig. 7. The accumulative reward during learning to compare methods

5   Discussions

The results show the feasibility of attention control in decision space. There are some general advantages for learning in this new space. The major ones are listed below:

- The local knowledge gained by different experts is utilized in a distributed manner by decision agents to make a unified and more confident decision. This is in fact the main justification behind any fusion algorithm.
- Decision agents share the decision space. So, their decisions can be verified to anticipate which decision agents are redundant, which decisions are more informative

---

[2] The number of episodes required to reach a perfect recognition rate.

and even which ones contain complementary information. It is obvious that there is no such information straightforward available in perceptual space. This information can be utilized to further reduce the learning time in decision space.

- By attention control in decision space, we can take advantage of diverse available types of learning methods for decision agents. In fact, each decision agent can use the most suitable learning method regardless of what methods the other ones employ. This benefit is gained because all agents share the decision space. Possibility of using different learning methods across decision agents enables the designer to use dissimilar types of information -such as training data, expert knowledge, etc- and sensors for training different decision agents.

- Another issue to discuss is the fact that, transferring the attention control learning from perceptual space to decision space results in learning decision fusion. Decision fusion has some major advantages (like reliability, robustness and survivability) not only because of fusion [13] but also due to its boosting characteristics. Schapire in [14] describes: "Boosting is a general method for improving the accuracy of any given learning algorithm. It refers to a general and provably effective method of producing a very accurate prediction rule by combining rough and moderately inaccurate rules of thumb." The reason behind the claim that our proposed structure for attention control in decision space implements boosting is that "while the performance of each local expert (decision agent) is less than or equal to chance, by using learning attention control we can improve the performance considerably." Despite the motioned general benefits, the proposed representation of the decision space seems not to be theoretically compact. This problem can be quantified through a simple order computation for the two approaches which comes in Table 4:

Table 4. Comparing Order of State-Action for both approaches

	In Decision Space			In Feature Space	
Parameters	M: Number of Motor Actions m: number of sensors f: discretization level in sensory space c: discretization level in decision space n: number of decision agents k: number of sensors observed by each decision agent				
Theoretical Order of States-Action	$M\,.(n\,f^{k} + c^{n(M-1)})$			$M.\,f^{m}$	
Example (Theoretical Number of State-Action)	$M = 9$	$m = n = 3$	$f = 4$	$c = 10$	$k = 1$
	10^{24}			576	
Practical Number of State-Action	$M.\,(n f^{k} + C)$ $C = number\ of\ sparse\ points$ $in\ decision\ space$			$M.\,f^{m}$	
	1008 with $C \le 100$			576	

Above computation theoretically shows that the number of states in decision space is very large and expresses state explosion. While, as tested in practice, the number of exiting states in decision space is very much fewer than $c^{n(M-1)}$. It means that the agent does not even go into most of the theoretically mentioned states. In other words, the space is considerably sparse. So, there is no need to reserve any space for non-existing states and be aware of their values; which results in reasonable learning speed. We are not sure if the mentioned sparseness is hold such strongly in all practical cases. Therefore, it is one of our main concerns to find a more compact representation for the decision space to preferably speed up the learning and become robust to missing information and noise. One solution is not quantizing the decision space and using continuous space RL methods [15].

6 Conclusions and Future Works

The proposed approaches are our primary steps taken to bold the main requirements of a general framework for learning attention control in a multi-modal as well as dynamic perceptual space during learning to perform a complex decision making task, such as autonomous driving which surely contains many different distracters. It is expected that if there were many distracters, the attention control algorithm would try to remove those irrelevant dimensions thus accelerate learning process considerably. The main outcome of the paper is to show that learning attention control is feasible in decision space and the results are comparable with those attained in the perceptual space. Learning attention control in decision space benefits some interesting advantages over learning attention control in perceptual space. The major ones are sharing the common space (decision space) among tiny decision agents, utilizing not necessary similar learning algorithms for decision agents and finally making a more confident decision. There are many extensions planned for the proposed approach and the most important one is finding a more compact and yet meaningful decision space to learn attention in it with preferably higher advantages such as faster learning speed, lower cost and maybe more robustness. Another extension is learning to expand the perceptual space in a gradual manner.

Acknowledgments. This research was supported by University of Tehran and has been realized in close collaboration with the BACS project supported by EC-contract number FP6-IST-02´140, Action line: Cognitive Systems.

References

1. James, W.: The principles of psychology. Holt, New York (1890)
2. Reichle, E., Laurent, A.: Using Reinforcement Learning to Understand the Emergence of "Intelligent" Eye-Movement Behavior During Reading. Psychological Review Copyright 2006 by the American Psychological Association 113(2), 390–408 (2006)
3. Shariatpanahi, H.F., Ahmadabadi, M.N.: Biologically Inspired Framework for Learning and Abstract Representation of Attention Control. In: Proceedings of International Workshop on Attention in Cognitive Systems, at IJCAI 2007, Hyderabad, India, pp. 63–80 (2007)

4. Fritz, G., Seifert, C., Paletta, L., Bischof, H.: Attentive object detection using an information theoretic saliency measure. In: Paletta, L., Tsotsos, J.K., Rome, E., Humphreys, G.W. (eds.) WAPCV 2004. LNCS, vol. 3368, pp. 29–41. Springer, Heidelberg (2005)
5. Paletta, L., Fritz, G., Seifert, C.: Cascaded Sequential Attention for Object Recognition with Informative Local Descriptors and Q-learning of Grouping Strategies. In: Proceedings of the IEEE Computer Society Conference on Computer Vision and Pattern Recognition (2005)
6. Gonc, L., Giraldi, G., Oliveira, A., Grupen, R.: Learning Policies for Attentional Control. In: IEEE International Symposium on Computational Intelligence in Robotics and Automation, pp. 294–299 (1999)
7. Darrell, T.: Reinforcement Learning of Active Recognition Behaviors. In: Portions of this paper previously appeared in Advances in Neural Information Processing Systems (NIPS 1995), vol. 8. MIT Press, Cambridge (1995); Vidyasagar M. (ed.) Intelligent Robotic Systems, pp. 73-80. Tata Press (1998)
8. Maunsell, J., Treue, S.: Feature-based attention in visual cortex. TRENDS in Neurosciences, TINS special issue: The Neural Substrates of Cognition 29(6) (2006)
9. Rizzolatti, G., Fogassi, L., Gallese, V.: Neurophysiological mechanisms underlying the understanding and imitation of action. Review Neuroscience (2001)
10. Rizzolatti, G., Riggio, L., Dascola, I., Umilta, C.: Reorienting Attention across the Horizontal and Vertical Meridians: Evidence in Favor of a pre-motor theory of attention. NeuroPhysiologia 25(1A), 31–40 (1987)
11. Putterman, M.L.: Markov Decision Processes. John Wiely and Sons, New York (1994)
12. Kuncheva, L., Bezdek, J., Duin, R.: Decision templates for multiple classifier fusion. The Journal of the Pattern Recognition Society 34(2), 299–314 (2001)
13. Zhu, Y.: Multi-sensor Decision and Estimation Fusion. The International Series on Asian Studies in Computer and Information Science 14 (2002) ISBN: 978-1-4020-7258-1
14. Schapire, R.: The Boosting Approach to Machine Learning: An Overview, AT&T Labs, Research Shannon Laboratory, Nonlinear Estimation and Classification. Springer, Heidelberg (2003)
15. Sutton, R.S., Barto, A.G.: Reinforcement Learning. MIT Press, Cambridge (1998)

Automated Visual Attention Manipulation

Tibor Bosse[1], Rianne van Lambalgen[1], Peter-Paul van Maanen[1,2], and Jan Treur[1]

[1] Vrije Universiteit Amsterdam, Department of Artificial Intelligence
De Boelelaan 1081a, 1081 HV Amsterdam, The Netherlands
{tbosse,rm.van.lambalgen,treur}@cs.vu.nl
[2] TNO Human Factors, P.O. Box 23, 3769 ZG Soesterberg, The Netherlands
peter-paul.vanmaanen@tno.nl

Abstract. In this paper a system for visual attention manipulation is introduced and formally described. This system is part of the design of a software agent that supports naval crew in her task to compile a tactical picture of the situation in the field. A case study is described in which the system is used to manipulate a human subject's attention. To this end the system includes a Theory of Mind about human attention and uses this to estimate the subject's current attention, and to determine how features of displayed objects have to be adjusted to make the attention shift in a desired direction. Manipulation of attention is done by adjusting illumination according to the calculated difference between a model describing the subject's attention and a model prescribing it.

1 Introduction

In the domain of naval warfare, it is crucial for the crew of the vessels involved to be aware of the situation in the field. Examples of important questions that should be addressed continuously are "in which direction are we heading?", "are we currently under attack?", "are there any friendly vessels around?", and so on. To assess such issues, one of the crew members is usually assigned the Tactical Picture Compilation Task (TPCT): the task to identify and classify all entities in the environment (e.g., [11]). This is done by monitoring a radar screen for radar contacts, and reasoning with the available information in order to determine the type and intent of the contacts on the screen. However, due to the complex and dynamic nature of the environment, this person has to deal with a large number of tasks in parallel. Often the radar contacts are simply too numerous and dynamic to be adequately monitored by a single human, which compromises the performance of the task.

For these reasons, it may be useful to offer the human some support from an intelligent ambient system, consisting of software agents that assist him in the execution of the Tactical Picture Compilation Task. For example, in case the human is directing its attention on the left part of a radar screen, but ignores an important contact that just entered the radar screen from the right, such an agent may alert him about the arrival of that new contact. To be able to provide this kind of intelligent support, the system somehow needs to maintain a model of the cognitive state of the human: in this case the human's focus of attention. It should have the capability to attribute mental, and in particular attentional (e.g., [12], [13], [14]) states to the

L. Paletta and J.K. Tsotsos (Eds.): WAPCV 2008, LNAI 5395, pp. 257–272, 2009.
© Springer-Verlag Berlin Heidelberg 2009

human, and to reason about these. In psychology and philosophy this characteristic is often referred to as Theory of Mind (or ToM, see, e.g., [1]). According to [7], agents, both human and software, can exploit a Theory of Mind for two purposes: to *anticipate* the behaviour of other agents (e.g., preparing for the consequences of certain actions that the other will probably perform), and to *manipulate* it (e.g., trying to influence the actions that the other will perform). In case of an intelligent system to support naval crew members, both purposes are relevant, but require a different type of support. This study is related to the latter type, the type that tries to manipulate the focus of attention.

A number of approaches in the literature address the development of software agents with a Theory of Mind; e.g., [16], [7]. Usually, such agents maintain, in one way or the other, a model of the epistemic (e.g., beliefs) and/or motivational states (e.g., desires, intentions) of other agents. However, for the situation sketched above, such agents ideally also have insight in another agent's attentional states. After all, if a supportive agent is to find out whether the human is ignoring some contact, it needs to have some knowledge about which contacts the person is paying attention to. This idea is in line with the theories of cognitive scientists like Gärdenfors [9], [10], who claims that humans have a Theory of Mind that is not only about beliefs, desires, and intentions, but also about other mental states like attentional, emotional, and awareness states.

The current paper is the result of a project that aims to develop intelligent agents to support naval crew members in the Tactical Picture Compilation Task, based on the ideas described above. To this end, four models have been developed. First, a dynamical model of human attention is needed, which estimates where the person's attention is, based on information about features of objects on the screen and the person's gaze. Second, a reasoning model is needed to reason through the first model in order to generate beliefs on attentional states at any point in time. Third, a model is needed that compares the output of the second model with some normative attention distribution, and determines whether there is a discrepancy. Finally, a model is needed that uses the output of the third model to determine how to alert the human that he is ignoring something important. An initial version of the first two models has already been developed and were adopted from this earlier work ([5], [6], respectively [2]). The current paper has its focus on the development of the other two models.

Section 2 presents a brief introduction of the existing literature on attentional processes, which helps to understand the choices made within this paper. Next, Section 3 formally describes the different models within the supportive software agent, and presents some simulations that were performed to test the behaviour of the model at a conceptual level. In Section 4, the whole approach is applied in a real-world a case study, using human gaze data and a tactical picture compilation task environment. Finally, Section 5 is a discussion.

2   Manipulation of Attention

Typically, a person's attention is influenced both by top-down and by bottom-up processes. The former means that observers orient their attention in a goal-directed manner, as a consequence of their expectations or intentions [19]. For example, when searching for a friend in the crowd, attention is guided top-down [20]. In contrast, the latter means that attention is elicited by a (highly salient) trigger from the environment. For example, one green circle among several blue circles will "pop-out" and attention will be directed to this object [22].

In this project the focus is primarily on adjusting the features of a specific location, such that only bottom-up attention is manipulated. Features that are mainly known to influence attention are intensity (luminance), colour and orientation. Previous research shows that attention can be elicited both by the contrast with stimuli at other locations [12], [15], [17] and the abrupt change of a feature, like luminance [21], [23] or form [23].

Several cognitive models on attention have been proposed and show that it is possible to predict attention allocation based on a saliency map, calculated from features of a stimulus, like luminance, colour and orientation [13], [18]. These models are not dynamic in the sense that they are based on existing information from the environment. However, if indeed the change of a specific feature (like luminance) can cause an attention shift in the human performing a task considered, a model can be used to realise this change. This way, humans who have to direct their attention to a large number of locations in parallel can be supported to adequately perform their task.

3 Formalisation of a Theory of Mind for Attention

In this section it is shown how the Theory of Mind for attention within the software agent was designed. First, in Section 3.1 the general setting is described, distinguishing four models. In subsequent subsections 3.2, 3.3, 3.4, and 3.5 these four models are described in more detail.

3.1 Overall Setting

A Theory of Mind enables an agent to analyze another agent's mind, and to act according to the outcomes of such an analysis and its own goals. For the general case such processes require some specific facilities.

(1) A *representation of a dynamical model* is needed describing the relationships between different mental states of the other agent. Such a model may be based on qualitative causal relations, but it may also concern a numerical dynamical system model that includes quantitative relationships between the other agent's mental states. In general such a model does not cover all possible mental states of the other agent, but focuses on certain aspects, for example on beliefs and desires, on emotional states, on the other agent's awareness states, or on attentional states as in this paper.

(2) Furthermore, *reasoning methods to generate beliefs on the other agent's mental state* are needed to draw conclusions based on the dynamical model in (1) and partial information about the other agent's mental states. This may concern deductive-style reasoning methods performing forms of simulation based on known inputs to predict certain output, but also abductive-style methods reasoning from output of the model to (possible) inputs that would explain such output.

(3) Moreover, when in one way or the other an estimation of the other agent's mental state has been found out, it has to be *assessed whether there are discrepancies*

between this state and the agent's own goals. Here also the agent's self-interest comes in the play. It is analyzed in how far the other agent's mental state is in line with the agent's own goals, or whether a serious threat exists that the other agent will act against the agent's own goals.

(4)   Finally a *decision reasoning model* is needed to decide how to act on the basis of all of this information. Two types of approaches are possible. A first approach is to take the other agent's state for granted and prepare for the consequences to compensate for them as far as these are in conflict with the agent's own goals, and to cash them as far as they can contribute to the agent's own goals (*anticipation*). For the navy case, an example of anticipation is when it is found out that the other agent has no attention for a dangerous object, and it is decided that another colleague or computer system will handle it (dynamic task reallocation). A second approach is not to take the other agent's mental state for granted but to decide to try to get it adjusted by affecting the other agent, in order to obtain a mental state of the other agent that is more in line with the agent's own goals (*manipulation*). This is the case addressed in this paper.

In this paper the general pattern sketched above is applied to the way in which a (software) agent can attempt to adjust the other (human) agent's attention, whenever required. To this end the software agent uses four types of facilities:

- *A dynamical model for attention*
  Representation of a dynamical attention model: a model that provides as output an estimation of the current attention distribution, based on input about features of objects on the screen and the other agent's gaze.
- *A reasoning model to generate beliefs about attentional states*
  These methods are used to estimate the attention given inputs about features of the objects and the other agent's gaze.
- *A discrepancy assessment model*
  This concerns a model to determine whether it is desirable that the attention distribution is changed, and to which extent: the discrepancy between actual and desirable attentional states
- *A decision reasoning model*
  This is a model to determine how, given a desire to adjust the attention distribution in certain respects, the inputs for the attention model have to be changed, to obtain an attention distribution as output, which is adjusted as desired.

The dynamical attention model is taken over from [5], [6] and is only briefly summarised below. The second model is kept relatively simple: beliefs on the attentional state are generated just based on (internal) simulation of the attention model. The third and fourth model form the most crucial part of this paper.

3.2 The Dynamical Attention Model Used

The attention distribution at *time* t is an assignment of attention values $AV(s, t)$ to a set of attention *spaces* s. Attention spaces are squares within a grid. The attention distribution is assumed to have a certain persistency. At each point in time the new attention is related to the previous attention, by

$$AV(s,t) = \lambda \cdot AV(s,t-1) + (1-\lambda) \cdot AV_{norm}(s,t)$$

where   is the decay parameter that results in the decay of the attention value of space s at time point $t-1$. Note that higher values for   result in a higher persistency and lower decay and vice versa. Here $AV_{norm}(s,t)$ is determined by a normalisation process keeping the total amount of attention fixed. This is described by:

$$AV_{norm}(s,t) = \frac{AV_{new}(s,t)}{\sum_{s'} AV_{new}(s',t)} \cdot A(t)$$

$$AV_{new}(s,t) = \frac{AV_{pot}(s,t)}{1 + \alpha \cdot r(s,t)^2}$$

Here $AV_{new}(s,t)$ is defined as follows. An important aspect of the visual attentional state is human gaze behaviour. Therefore the relative distance of each space s to the gaze point (the centre) is an important factor in determining the attention value of s. Mathematically this is modelled by the formula above, where $AV_{pot}(s,t)$ is the potential attention value of s at time point t. The term r(s,t) is taken as the Euclidian distance between the current gaze point and s at time point t (multiplied by an importance factor α which determines the relative impact of the distance to the gaze point on the attentional state, which can be different per individual and situation):

$$r(s,t) = d_{eucl}(gaze(t),s)$$

$$AVpot(s,t) = \sum_{maps\ M} M(s,t) \cdot w_M(s,t)$$

Here the potential attention value $AV_{pot}(s,t)$ is calculated as follows, based on the properties of the space (i.e., of the types of objects present) at that time (for instance features such as colour, intensity, and orientation contrast, amount of movement). For each of such a feature a specific *saliency map* describes its potency of drawing attention; e.g., [8], [13], [14] Because not all features are equally highlighting, an additional weight for every map is used. Formally the above can be described as shown above, where for any feature there is a saliency map M, for which M(s,t) is the unweighted potential attention value of s at time point t, and $w_M(s,t)$ is the weight for saliency map M, where $1 \le M(s,t)$ and $0 \le w_M(s,t) \le 1$. The exact values for the weights depend on the specific application.

3.3   Reasoning Model to Generate Beliefs About Attentional States

The reasoning method to generate beliefs about attentional states is kept simple. The model described in Section 3.2 as a dynamical system model (based on a difference equation) is just used by the software agent as an internal simulation model to generate new attentional states out of the previous ones and information about the current features of the objects. This is done by a forward reasoning method (forward in time) as described in [2]. This reasoning method can be used to make predictions on future states, or on making an estimation of the current state based on information acquired in the past. This reasoning method occurs in the literature in many variants, in different contexts and under different names, varying from, for example, computational (numerical) simulation based on difference or differential equations, qualitative simulation, causal reasoning, execution of executable temporal logic formulae, and forward chaining in rule-based reasoning, to generation of traces by transition systems and finite automata. The basic specification of this reasoning model can be expressed as follows, where belief(leads_to_after(I, J, D)) (the belief that when state I holds at time T, then J will hold after time duration D) is used as the agent's internal representation format for dynamical system models, and belief(at(I, T)) as a representation of its information on the world (including human processes) at different points in time; moreover, → means that when the antecedent holds, the consequent will follow. Here, I and J are predicates that may represent world states like 'AV(s,t) = 0.5' or 'luminance value 0.8 is assigned to s'.

Belief Generation based on Positive Forward Simulation
If it is believed that I holds at T and that I leads to J after duration D, then it is believed that J holds after D.

$$belief(at(I, T)) \land belief(leads\_to\_after(I, J, D)) \rightarrow belief(at(J, T+D))$$

If it is believed that I1 holds at T and that I2 holds at T, then it is believed that I1 and I2 holds at T.

$$belief(at(X1,T)) \land belief(at(X2, T)) \rightarrow belief(at(and(X1, X2), T))$$

This is done by calculations following the formulae described above.

3.4   A Model to Determine Discrepancy between Actual and Desirable Attention

The discrepancy between actual and desirable attention can be determined as soon as a model is available for what the desirable attention distribution is (sometimes called a prescriptive model). For the case addressed in this paper this means that in a computational manner it is assessed which objects deserve attention, an assessment on the basis of features such as distance, speed and direction of an object[1]. In fact, this is close to the first part of the task the human is performing: identification of the relevant objects to be handled.

3.5   A Decision Model for Attention Adjustment

The model for adjustment of the attention distribution has as input the discrepancy determined by the model described in Section 3.4, and also makes use of the explicitly represented dynamical model as described in Section 3.2. The general idea is that the relations between variables within this model are followed in a backward manner, thereby propagating the desired adjustment from the attentional state variable to the features of the object at the screen. The general pattern behind this operation on a dynamical model representation is illustrated in Figure 1. Here v_l is the (desired)

---

[1] It is assumed that the agent has tactical domain knowledge that enables it to make such assessments.

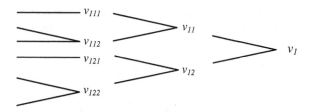

Fig. 1. Dependencies between variables in a dynamical system model

output of a model, and by branches the variables on which this depends are depicted, until the leaves where actual adjustments can be made[2].

This is a form of *desire refinement*: starting from the root variable, by a step-by-step process a desire on adjusting a parent variable is refined to desires on adjustments of the children variables, until the leave variables are reached. The starting point is the desire on the root variable, which is the desired adjustment of the attentional state; this is determined by.

belief(av(s)<h) ∧ desire(a(v)>h) ∧ belief(has_value(av(s), v)) → desire(adjust_by(av(s), (h-v)/v)

Note that here the adjustment is taken relative (expressed by division of the difference h-v by v). Suppose as a point of departure (given the discrepancy assessment) an adjustment Δv_1 is desired, and that v_1 depends on two variables v_{11} and v_{12} that are adjustable (the non-adjustable variables can be left out of consideration). Then by elementary calculus as a linear approximation the following relations between required adjustments can be obtained:

$$\Delta v_1 = \frac{\partial v1}{\partial v11} \Delta v_{11} + \frac{\partial v1}{\partial v12} \Delta v_{12}$$

This formula is used to determine the desired adjustments Δv_{11} and Δv_{12}, where by weight factors μ_{11} and μ_{12} the proportion can be indicated in which the variables should contribute to the adjustment: $\Delta v_{11}/\Delta v_{12} = \mu_{11}/\mu_{12}$.

$$\Delta v_1 = \frac{\partial v1}{\partial v11} \Delta v_{12}\mu_{11}/\mu_{12} + \frac{\partial v1}{\partial v12}\Delta v_{12} = \left(\frac{\partial v1}{\partial v11} \mu_{11}/\mu_{12} + \frac{\partial v1}{\partial v12} \right) \Delta v_{12}$$

So the adjustments can be made as follows:

$$\Delta v_{12} = \frac{\Delta v1}{\frac{\partial v1}{\partial v11}\mu11/\mu12 + \frac{\partial v1}{\partial v12}}$$

$$\Delta v_{11} = \mu_{11}/\mu_{12} \frac{\Delta v1}{\frac{\partial v1}{\partial v11}\mu11/\mu12 + \frac{\partial v1}{\partial v12}} = \frac{\Delta v1}{\frac{\partial v1}{\partial v11} + \frac{\partial v1}{\partial v12}\mu12/\mu11}$$

Special cases are $\mu_{11} = \mu_{12} = 1$ (*absolute equal contribution*) or $\mu_{11} = v_{11}$ and $\mu_{12} = v_{12}$ (*relative equal contribution*: in proportion with their absolute values). As an

---

[2] For the moment, deterministic relationships between variables are assumed. However, in a later stage, the agent might learn such relationships.

example, consider a variable that is just the weighted sum of two other variables (as is the case, for example, for the aggregation of the effects of the features of the objects on the attentional state):

$$v_1 = w_{11}v_{11} + w_{12}v_{12}$$

For this case

$$\frac{\partial v1}{\partial v11} = w_{11} \qquad\qquad \frac{\partial v1}{\partial v12} = w_{12}$$

and

$$\Delta v_{11} = \frac{\Delta v1}{w11 + w12\ \mu12/\mu11} \qquad \Delta v_{12} = \frac{\Delta v1}{w11\ \mu11/\mu12 + w12}$$

For example when $\mu_{11} = \mu_{12} = 1$ this results in

$$\Delta v_{11} = \frac{\Delta v1}{w11 + w12} \qquad \Delta v_{12} = \frac{\Delta v1}{w11 + w12}$$

Assuming $w_{11} + w_{12} = 1$ in addition, this results in $\Delta v_{11} = \Delta v_{12} = \Delta v_1$

Another setting, which actually has been used in the model is to take $\mu_{11} = v_{11}$ and $\mu_{12} = v_{12}$. In this case the adjustments are assigned proportionally; for example, when v_1 has to be adjusted by 5%, also the other two variables on which it depends need to contribute an adjustment of 5%. Thus the relative adjustment remains the same through propagations:

$$\frac{\Delta v11}{v11} = \frac{\Delta v1}{w11 + w12\ v12/v11}/v_{11} = \frac{\Delta v1}{w11v11 + w12\ v12} = \frac{\Delta v1}{v1}$$

This shows the general approach on how desired adjustments can be propagated in a backward manner through a dynamical model. Thus a desired adjustment of the attentional state as output at some point in time can be related to adjustments in the features of the displayed objects as inputs at previous points in time. For the case study undertaken this approach has been applied, although at some points in a simplified form. One of the simplifications made is that due to the linearity of most dependencies in the model, adjustments have been used that just propagate without any modification. An example of a rule specified to achieve this propagation process is:

desire(adjust_by(u1, a)) ∧ belief(depends_on(u1, u2)) ⇸ desire(adjust_by(u2, a))

Here the adjustments are taken relative, so, this rule is based on $\Delta u_2 / u_2 = \Delta u_1 / u_1$ as derived above for the linear case. When at the end the leaves are reached, which is represented by the belief that they are directly adjustable, then from the desire an intention to adjust them is derived.

desire(adjust_by(u, a)) ∧ belief(directly_adjustable(u)) ⇸ intention(adjust_by(u, a))

If an intention to adjust a variable u by a exists with current value b, the new value b+ α*a*b to be assigned to u is determined; here α is a parameter that allows the modeler to tune the speed of adjustment:

intention(adjust_by(u, a)) ∧ belief(has_value_for(u, b)) →
        performed(assign_new_value_for(u, b+ α*a*b))

This rule is applied for variables that describe features f of objects at locations s, i.e., instances for u of the form feature(s, f). Note that each time the adjustment is propagated as a value relative to the overall value.

3.6 Simulation Results

To test whether the approach described above yields the expected behaviour, it has been used to perform a number of simulation experiments in the LEADSTO simulation environment [4]. This environment takes a specification of causal relationships (in the format as shown in the previous sections) as input, and uses this to generate simulation traces. The simulations shown here address a slightly simplified case, where the radar screen has been split up in 4 locations. For the time being, it is assumed that each location contains one contact, and that these contacts stay within their locations. The features of the contacts that are manipulated are luminance, size, and level of flashing. Initially, each contact starts with the same features, but during the simulation these features are manipulated, based on the prescribed (or desired) attention. This desired attention is generated randomly, where every 50 time units a next location is selected where the attention should be. Furthermore, the behaviour of the human gaze is generated as follows: after each adaptation of the features, the gaze moves to one of the four locations, with a probability that is proportional to the saliency of the contact at that location.

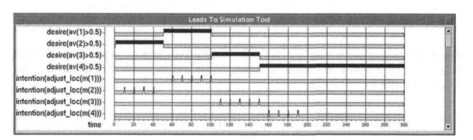

Fig. 2. Model-based reasoning process. First it is intended (several times) to adjust a feature value at location 2, then at location 1, then at location 3, and finally at location 4.

The results of an example simulation run are depicted in Figures 2 to 5. In these figures, time is on the horizontal axis, and the different state of the process is shown in the vertical axis. A dark line indicates that a state is true at a certain time point. Note that some information has been omitted due to space limitations. Figure 2 shows the model-based reasoning process of the agent, in terms of desires and intentions. Figures 3, 4, and 5 show, respectively, the estimated attention, the human's gaze, and the value of the feature "luminance" at different locations over time. As shown in Figure 2, initially it is desired that at least 50% of the human's attention is at location 2 (desire(av(2)>0.5)). Since this is not the case (see Figure 3), the luminance of the

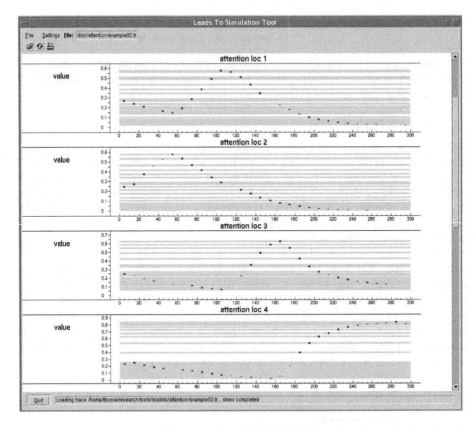

Fig. 3. Estimated attention at different locations. Initially the highest attention value is estimated to be at location 2 (with a peak around time point 55), then at location 1, then at location 3, and finally at location 4.

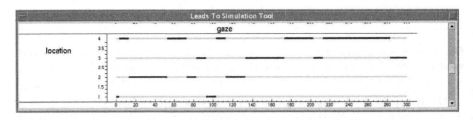

Fig. 4. Dynamics of gaze. The vertical axis denotes the location of the gaze, which switches between location 1, 2, 3, and 4.

contact at location 2 is increased (see Figure 5). As a result, the human's gaze shifts towards this location (see Figure 4), which increases his attention for location 2. In the rest of the simulation, this pattern is repeated for different locations.

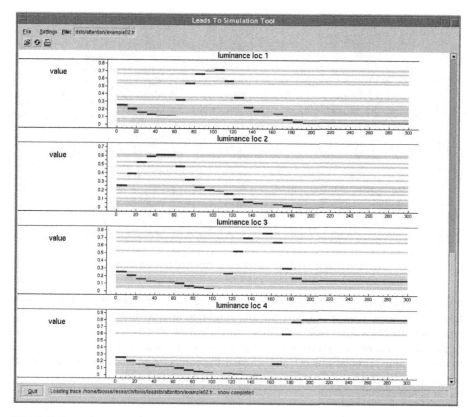

Fig. 5. Values of feature 'luminance' at different locations. First the luminance at location 2 is increased, then at location 1, 3, and 4 (note that values are normalised).

4 Case Study

To test the approach in a real world situation, and obtain an initial validation, a case study with human subjects while executing the Tactical Picture Compilation Task was undertaken. In Section 4.1 the environment is explained, Section 4.2 discusses some implementation details of the system tailored to the environment, and in Section 4.3 the first results are discussed.

4.1 Environment

The task used for this case study is an altered version of the identification task described in [11] that has to be executed in order to build up a tactical picture of the situation, i.e. the Tactical Picture Compilation Task (TPCT). The implementation of the software is done in Gamemaker [25]. In Figure 6 a snapshot of the interface of the task environment is shown. The goal is to identify the five most threatening contacts (ships). In order to do this, participants have to monitor a radar display where contacts in the surrounding areas are displayed. To determine if a contact is a possible threat,

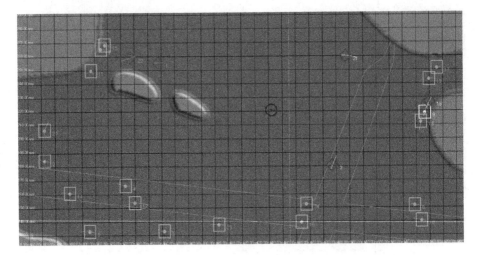

Fig. 6. A snapshot of the interface of the used task environment

different criteria have to be used. These criteria are the identification criteria (idcrits) that are also used in naval warfare, but are simplified in order to let naive participants learn them more easily. These simplified criteria are the speed (depicted by the length of the tail of a contact), direction (pointer in front of a contact), distance of a contact to the own ship (circular object), and whether the contact is in a sea lane or not (in or out the large open cross). Contacts can be identified as either a threat (diamond) or no threat (square).

4.2 Implementation

The system is further developed and evaluated using Matlab. The output of the environment described in Section 4.1 was used and consisted of a representation of all properties of the contacts visible on the screen, i.e. speed, direction, if it is in a sea lane or not, distance to the own ship, location on the screen and contact number. In addition, data from a Tobii x50 eye-tracker [24] were retrieved from a participant executing the TPC task. All data were retrieved several times per second and were used as input for the system. Once the system was tailored to the TPC case study, the eventual implementation of it was done in C#. The output of the implementation of the system causes the saliency of the different objects on the screen to either increase or decrease, which may result in a shift of the participant's visual attention. As a result, the participant's attention is continuously manipulated in such a way that it pays attention to the objects that are considered relevant by the system. The results of this implementation are described below.

4.3 Results

The first results of the implemented system are best described by a number of example snapshots of the outcomes of the system in three different situations over time (see Figure 7).

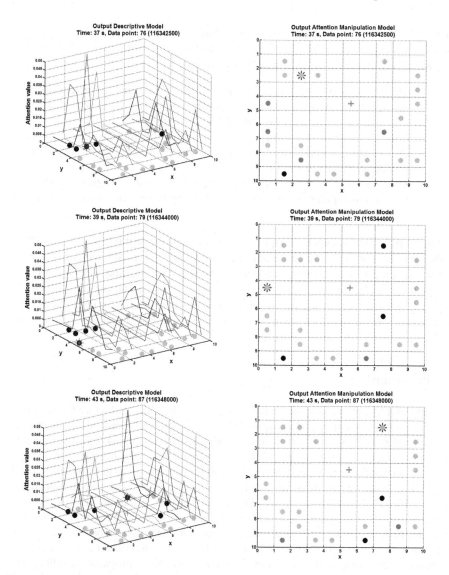

Fig. 7. Estimation of the participant's attention division (left figures) and reaction of the system (right figures) in three different situations

On the left side of Figure 7 the darker dots correspond to the system's estimation of those contacts to which the participant is paying attention. On the right side of the figure, the darker dots correspond to those contacts where attention manipulation is initiated by the system (in this case, by increasing its saliency). On both sides of the figure a cross corresponds to the own ship, a star corresponds to the eye point of gaze, and the x- and y-axes represent the coordinates on the interface of the TPCT. In the pictures to the left, the z-axis represents the estimated amount of attention. The darker dots on the left side are a result of the exceedance of this estimation of a certain threshold (in this case .03). Thus, a peak indicates that it is estimated that the participant has attention for that location. Furthermore, from top to bottom, the following three situations are displayed in Figure 7:

1. After 37 seconds since the beginning of the experiment, the participant is not paying attention to region A at coordinates (7.5,1.5), while no attention manipulation for region A is initiated by the system.
2. After 39 seconds, the participant is not paying attention to region A, while the attention should be allocated to region A, and therefore attention manipulation for region A is initiated by the system.
3. After 43 seconds, the participant is paying attention to region A, while no attention manipulation for region A is done by the system, because this is not needed anymore.

The output of the attention manipulation system and the resulting reaction in terms of the allocation of the participant's attention in the above three situations, show what one would expect of an accurate system of attention manipulation. As shown in the two pictures at the bottom of Figure 7, in this case the agent indeed succeeds in attracting the attention of the participant: both the gaze (the star in the bottom right picture) and the estimated attention (the peak in the bottom left picture) shift towards the location that has been manipulated.

5 Discussion

An important task in the domain of naval warfare is the Tactical Picture Compilation Task: the task to identify and classify all entities in the environment and determine the consequences in terms of tactical possibilities and constraints. However, due to the complex and dynamic nature of the environment, it is very difficult for a single human to perform this task adequately. Therefore, the current paper proposes to offer the person some support from an intelligent software agent that assists him in the Tactical Picture Compilation Task (TPCT). To this end, a number of models have been developed: 1) a dynamical system model for attention, 2) a reasoning model to generate beliefs about attentional states using the attention model for forward simulation, 3) a discrepancy assessment model, and 4) a decision reasoning model, again using the attention model, this time for backward desire propagation. This paper presented an initial version of such a supporting agent, focusing especially on the last two models, where the first two were adopted from earlier work in [5], [6], and [2]. This software agent is an example of a model-based intelligent ambient agent, as described in [3]. Within this type of agent an explicitly represented (dynamical system) model of human functioning plays an important role, for the case considered here the model of the human's attention. Such a model forms a basis for the application of dedicated model-based reasoning methods, as was also illustrated here: forward simulation reasoning and backward desire propagation reasoning.

After testing the system at a conceptual level by simulation, it has been implemented in a case study where participants have to perform a simplified version of the TPCT. Although no elaborated experimental validation has been performed as yet, initial results indicate that the agent is indeed able to adapt the features of objects in such a way that they attract the human's attention if necessary.

Concerning future work, an important challenge would be to perform a more elaborated validation of the supportive system. This can be done is several steps. First, to obtain more data, the experiment introduced in this paper will be performed with a larger number of participants. The resulting data can then be used to check (possibly using automated analysis tools) whether the supporting agent is successful in various situations. As part of this validation, also different strategies and parameter settings will be tested. For example, does adapting the shape of an object provide better results than adapting its luminance, or adapting multiple features? Similarly, in addition to manipulation of bottom-up attention, is it useful to manipulate top-down attention as well? Furthermore, in a later stage of the project, it is planned to evaluate whether the software agent indeed improves the task performance of the user.

Acknowledgments

This research was partly funded by the Royal Netherlands Navy (program number V524).

References

[1] Baron-Cohen, S.: Mindblindness: an essay on autism and theory of mind. MIT Press, Cambridge (1995)

[2] Bosse, T., Both, F., Gerritsen, C., Hoogendoorn, M., Treur, J.: Model-Based Reasoning Methods within an Ambient Intelligent Agent Model. In: Mühlhäuser, M., Ferscha, A., Aitenbichler, E. (eds.) AmI 2007 Workshops. CCIS 11, pp. 352–370. Springer, Heidelberg (2008)

[3] Bosse, T., Hoogendoorn, M., Klein, M., Treur, J.: An Agent-Based Generic Model for Human-Like Ambience. In: Mühlhäuser, M., Ferscha, A., Aitenbichler, E. (eds.) AmI 2007 Workshops. CCIS 11, pp. 93–103. Springer, Heidelberg (2008)

[4] Bosse, T., Jonker, C.M., van der Meij, L., Treur, J.: A Language and Environment for Analysis of Dynamics by Simulation. International Journal of Artificial Intelligence Tools 16(3), 435–464 (2007)

[5] Bosse, T., van Maanen, P.-P., Treur, J.: A Cognitive Model for Visual Attention and its Application. In: Nishida, T. (ed.) Proceedings of the 2006 IEEE/WIC/ACM International Conference on Intelligent Agent Technology (IAT 2006), pp. 255–262. IEEE Computer Society Press, Hong Kong (2006)

[6] Bosse, T., van Maanen, P.-P., Treur, J.: Simulation and formal analysis of visual attention in cognitive systems. In: Paletta, L., Rome, E. (eds.) WAPCV 2007. LNCS (LNAI), vol. 4840, pp. 463–480. Springer, Heidelberg (2007)

[7] Bosse, T., Memon, Z.A., Treur, J.: A Two-Level BDI-Agent Model for Theory of Mind and its Use in Social Manipulation. In: Proceedings of the AISB 2007 Workshop on Mindful Environments, pp. 335–342 (2007)

[8] Chen, L.Q., Xie, X., Fan, X., Ma, W.Y., Zhang, H.J., Zhou, H.Q.: A visual attention model for adapting images on small displays. ACM Multimedia Systems Journal (2003)

[9] Gärdenfors, P.: Slicing the Theory of Mind. In: Danish yearbook for philosophy, vol. 36, pp. 7–34. Museum Tusculanum Press (2001)

[10] Gärdenfors, P.: How Homo Became Sapiens: On the Evolution of Thinking. Oxford University Press, Oxford (2003)

[11] Heuvelink, A., Both, F.: Boa: A cognitive tactical picture compilation agent. In: Proceedings of the 2007 IEEE/WIC/ACM International Conference on Intelligent Agent Technology (IAT 2007). IEEE Computer Society Press, Los Alamitos (2007) (forthcoming)

[12] Itti, L., Koch, C.: A saliency-based search mechanism for overt and covert shifts of visual attention. Vision Research 40, 1489–1506 (2000)

[13] Itti, L., Koch, C.: Computational Modeling of Visual Attention. Nature Reviews Neuroscience 2(3), 194–203 (2001)

[14] Itti, L., Koch, U., Niebur, E.: A model of saliency-based visual attention for rapid scene analysis. IEEE Transactions on Pattern Analysis and Machine Intelligence 20, 1254–1259 (1998)

[15] Levitt, J.B., Lund, J.S.: Contrast dependence of contextual effects in primate visual cortex. Nature 387, 73–76 (1997)

[16] Marsella, S.C., Pynadath, D.V., Read, S.J.: PsychSim: Agent-based modeling of social interaction and influence. In: Lovett, M., Schunn, C.D., Lebiere, C., Munro, P. (eds.) Proc. of the Int. Conference on Cognitive Modeling, ICCM 2004, pp. 243–248 (2004)

[17] Nothdurft, H.: Salience from feature contrast: additivity across dimensions. Vision Research 40, 1183–1201 (2000)

[18] Parkurst, D., Law, K., Niebur, E.: Modeling the role of salience in the allocation of overt visual attention. Vision Research 42, 107–123 (2002)

[19] Posner, M.E.: Orienting of attention. Q. J. Exp. Psychol. 32, 3–25 (1980)

[20] Theeuwes, J.: Endogenous and exogenous control of visual selection. Perception 23, 429–440 (1994)

[21] Theeuwes, J.: Abrupt luminance change pops out; abrupt color change does not. Perception & Psychophysics 57(5), 637–644 (1995)

[22] Treisman, A.: Features and objects: The fourteenth Bartlett memorial lecture. Q. J. Experimental Psychology A 40, 201–237

[23] Turrato, M., Galfano, G.: Color, form and luminance caputer attention in visual search. Vision Research 40, 1639–1643 (2000)

[24] http://www.tobii.se

[25] http://www.yoyogames.com/gamemaker

Author Index

Allen, Harriet   124
Araabi, Babak N.   242
Aziz, Muhammad Zaheer   227

Bandera, Antonio   27
Belardinelli, Anna   112
Böhme, Christoph   41
Bosse, Tibor   257
Bruce, Neil D.B.   98

Carbone, Andrea   112
Cerf, Moran   15

Einhäuser, Wolfgang   15
Enescu, Valentin   166

Geerinck, Thomas   166, 197
Gobet, Fernand   183

Harel, Jonathan   1, 15
Heinke, Dietmar   41
Henderickx, David   166, 197
Höning, Nicolas   153
Humphreys, Glyn   124
Hussain, Fehmida   139
Huth, Alex   15

Koch, Christof   1, 15
König, Peter   153

Lane, Peter C.R.   183
Leonardis, Aleš   54

Maetens, Kathleen   197
Mancas, Matei   212
Marfil, Rebecca   27

Mavritsaki, Eirini   124
Mertsching, Bärbel   227
Mirian, Maryam S.   242

Nili Ahmadabadi, Majid   242

Perko, Roland   54
Pirri, Fiora   112

Rodríguez, Juan Antonio   27

Sahli, Hichem   166
Sandoval, Francisco   27
Schankin, Andrea   69
Schiele, Bernt   54
Schubö, Anna   69
Siegwart, Ronald R.   242
Smith, Richard Ll.   183
Soetens, Eric   197
Steger, Johannes   153

Templeman, Emma   85
Treur, Jan   257
Tsotsos, John K.   98

Underwood, Geoffrey   85
Underwood, Jean   85

Vanhamel, Iris   166
van Lambalgen, Rianne   257
van Maanen, Peter-Paul   257

Wilming, Niklas   153
Wojek, Christian   54
Wolfsteller, Felix   153
Wood, Sharon   139